Ocean

Asha de Vos

Ocean

From the Shore to the Abyss

with over 400 illustrations

Foreword by
Peter Godfrey-Smith

Contributors
Anthony J. Martin
Rebecca Helm
Helen Scales
Andrew Thaler

T&H

Contents

Anthony J. Martin

Rebecca Helm

Asha de Vos

Foreword

By Peter Godfrey-Smith

For animals of all kinds, the ocean was our first home. Life itself may have begun in the sea – that is still unclear – but there is little doubt that animals, our own sprawling branch on the total tree of life, evolved there. A marine ancestry is visible in our bodies and our genes.

The animal adventure in evolution began when some ancient collections of cells began working together in new ways. At first, these novel beings were probably no more than vague, rag-tag clumps, living on drifting bacteria and organic debris. When they began to take on a more definite form, we don't know what they looked like – perhaps like sponges, anchored on the seafloor, perhaps like filmy jellyfish, up in the water column. But all animals, from bees to cheetahs, have this soft-bodied oceanic origin.

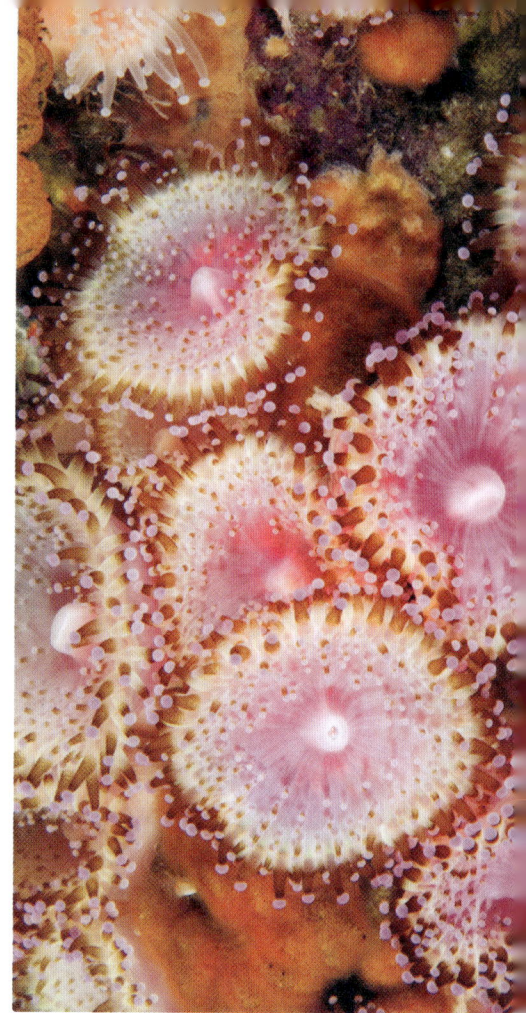

01

Life on land may have made the ocean seem foreign to us in some ways. But we are going home whenever we return to it.

This adventure in evolution diversified, became many adventures, sprouting new forms and lifestyles. Corals, anemones, nudibranchs, fish ... When we look through the animals in this book, we are looking, in every case, at our relatives, some nearer and others more distant. A decorator crab attaches stinging anemones to its body to wear as protection against predators such as octopuses. The crab, anemones and octopuses are all your distant cousins. As animals took on these diverse forms, they evolved action, vision, learning, remembering and navigation. Nervous systems made it possible for animals to swim and crawl, also to feel and remember. The mind evolved in the sea.

Later, several animal groups, including an odd-looking group of fish, began a move into a different sort of life, crawling up onto land. There, in blazing sunlight and soon surrounded by forests, animals began to change again. New kinds of bodies and minds evolved, including, eventually, our human minds. All living beings affect their environments; life does not leave things as they were. But human minds, with their effects amplified by social coordination, soon began a different sort of transformation of the world.

02

03

Human action and technology have a glorious side, but the expansion of our powers has led to us doing great harm to the oceans. Overfishing, mining and pollution are having grim effects. All these effects are less noticeable than comparable harm would be on land, because so much of it is hidden beneath the surface. But people around the world are now recognizing the damage being done. Some, in traditional coastal communities and indigenous societies, have been saying this for a long time. We need to look at the sea differently.

Life on land may have made the ocean seem foreign to us in some ways. But we are going home whenever we return to it. This is true of the smallest excursions. We are feeling the borderlands of those old haunts when we walk in the shallows along a beach, shells beneath our feet. We are looking into our origins when we float above a coral reef with a mask and snorkel. We are revisiting our past when we descend into the blue with scuba gear, observed by fish. I imagine the fish, octopuses, and dolphins asking: 'Are you sure you're glad you left?'

As this book shows, we should protect and appreciate, feel kinship and pride towards, our strange and wonderful first home.

7

Introduction

The ocean is a vast, three-dimensional space that makes up almost two-thirds of our planet. For those who do not have the opportunity to explore below the waves, it may seem to be merely a giant tank of water that spills endlessly beyond the horizon. But the ocean is not an empty expanse. It is teeming with life, home to some of the smallest animals on Earth and the very largest – the blue whale. Within the cracks and crevices of the myriad natural and humanmade marine habitats, all species great and small seek shelter and food, spend time with family and through the hubbub of their daily lives help keep this huge ecosystem, the beating heart of our planet, alive.

01

01 – A blacktip reef shark
hunting sardines
Carcharhinus melanopterus,
Sardina pilchardus

02 – Nudibranch
Hypselodoris maritima

03 – Peacock mantis shrimp
Odontodactylus scyllarus
Ari Atoll, Indian Ocean,
The Maldives

04 – Two cleaner wrasses on
a lunar-tailed bigeye
Labroides dimidiatus,
Priacanthus hamrur

02

03

04

It is not just living things that drive the health of our oceans. Currents twist and turn around imaginary corners, shifting water thousands of kilometres to maintain reasonable temperatures – both inside the sea and in the Earth's atmosphere. Upwellings bring nutrient-rich water to the surface, providing food and nourishment to hungry plants and animals. Moving water masses change the properties of the ocean in some places more regularly than others, drawing in different species throughout the year. These intricate and inter-connected systems, along with a wide array of other physical oceanographic processes that occur beneath the waves, are critical to sustaining our seas. All of this keeps us alive, too.

Despite this, the ocean has historically borne the brunt of escalating human activities. Our growing needs have increased our demands on the ocean and its resources. In many ways, we have taken advantage of its generosity. We have extracted, fished, travelled, farmed, spilled, hunted, mined, transported – with barely a thought to look after our seas and return the favour. While the oceans are full of life, what we see today is just a fraction of what existed in the past. The shifting baseline leaves us unable to recognize how much impact we've already had and unaware of what that impact, in turn, means for us today. Most worryingly, it contributes to an alarming complacency about the future.

With all that in mind, this book is a love letter to the ocean. It is an attempt to shine a light on some of the most exciting aspects of the deep blue, from tiny creatures to marine giants, dark trenches to colourful corals, rusting shipwrecks to cutting-edge marine-inspired inventions. Make no mistake, the ocean offers so much more than we could ever contain within the bindings of a book, so we have carefully hand-picked the stories we hope will inspire you and make you want to pass them on.

Ocean takes readers on a grand adventure through the layers of our resplendent seas. The various zones are divided by invisible lines, defined by changing characteristics associated with increasing depth – pressure, temperature and light. These physical properties determine the lives and lifestyles of the creatures that live within them.

We start on the shore. Today, it is home to mushrooming coastal development, intense human activity and built infrastructure that supports the world's oldest industry – fishing. This zone is the one used most by humans, but it is also the least explored ecologically. The place where the sea meets the land, our shores also provide portals into the past, where fossils, ripple marks and geological layers reveal how our planet might once have looked. Next we shift to the water's surface, another oft-overlooked zone. Despite its visibility, we all too often bypass the skin of the ocean in search of what lies beneath. But there are some truly remarkable creatures that flourish and breed here – fish that fly, rafting species hitching rides and weird and wonderful slugs, snails and jellies. Descending into the sunlight zone below, we witness an abundance of life that thrives under the rays of the sun and the light it shines. The zone that humans are best adapted to exploring, it reveals dazzling sweeps of colourful corals, gluts of fish gathering to spawn or make treacherous journeys across the ocean and rich ecosystems that connect micro-organisms, fish and large marine mammals in complex webs. Then, as we move further down into the twilight and midnight zones, we unearth the ingenious ways that a host of curious species navigate an area of darkness

We have extracted, fished, travelled, farmed, spilled, hunted, mined, transported – with barely a thought to look after our seas and return the favour.

– how they find fuel, avoid predators, seek out prey and locate safe refuges in which to hide. Finally, we reach the bottom: the abyss. This ocean layer demands creatures adapt in unusual ways to survive; it hosts some of nature's most highly specialized – and oddest-looking – beasts. The ocean floor is also the final resting place of numerous ill-fated ships, wartime submarines and even abandoned spacecraft, making the abyss a watery record, a museum, if you will, of fascinating cultural heritage and history.

As we reach our voyage's end, we want you to be inspired, and firm ambassadors for our ocean. But we also want to foster a deep connection with our seas through stories about 'the oceans and us' – our final chapter. Here, we look at how humans impact our seas, but mostly how we benefit from them over and over again. As a species, we have been shaped by our dependency on the ocean, while some of us have formed relationships with the sea – and the creatures that call it home – that defy expected boundaries.

Ultimately, as we conclude our homage to the ocean, we hope that hearts beat harder at the thought of the ocean – and even more so at the sight of it. That brains whirr in an effort to understand it; not to conquer it, but to protect it.

Humpback whales
Megaptera novaeangliae
A mother, calf and escort
near Hawaii, USA

The Shore

Introduction

Hundreds of thousands of years ago, when humans began travelling from the inland grasslands of Africa to other continents, their journeys would have brought them to the ocean for the first time. Seeing vistas with seemingly limitless water must have felt especially strange for people who had just traversed such arid landscapes, hosting few ponds or streams. The smacking of waves on the shore, as well as the twice-daily flooding and recession of tides that alternately inundated and exposed the earth and life below, probably inspired a similar sense of curiosity and awe. The briny organic scent and salty taste would have identified this water as something different from what they previously knew. This cognitive awakening to oceans was accompanied by encounters with sediments, rocks, plants and animals influenced or otherwise shaped by their marine environment.

For some ocean shores, their abundant resources and compelling sensory and aesthetic wonders gave ancient travellers good reason to explore and learn more about them; to use their resources; or to settle down so they could live with – and off – their riches every day. Eventually, some were brave enough to leave the shore and traverse the seas, only to set foot on other distant lands that became new homes.

Shores are not only places where horizon-spanning waters fringe the land, but also frontiers where sea- and land-dwelling creatures meet, often with shared evolutionary histories. Moreover, the lineages of all modern land plants and animals began in our oceans, and at least a few of those same groups – such as marine grasses and mammals – have returned. Over geological time, material changes to the makeup of our planet have occurred, leaving imprints for us to discern. Ancient shores, now far inland, have left evidence of their arrivals and departures in the limestone that bears mollusc shells and other fossils of shallow-marine pasts. These legacies are a reminder that modern shores are part of an interplay between Earth, moon, sun and life that has existed for and developed over billions of years and will continue well past our species' time.

Shores are not borders, but more like permeable boundaries with regular exchanges between marine and terrestrial realms. Shore positions also constantly shift daily with waves, tides and other such factors, or move more broadly with sea-level rises or falls. Although intertidal zones, such as tidal flats and beaches, are the strips of land most affected by waves and tides, ecosystems beside them – such as coastal dunes and maritime forests – are also influenced by sea spray and other oceanic input. Similarly, shallowly submerged areas and their life just offshore are tied to sunlight from above, as well as to nutrients and sediments that come from the land.

Shores are not only places where horizon-spanning waters fringe the land, but also frontiers where sea- and land-dwelling creatures meet, often with shared evolutionary histories.

These shallows on continental margins or around islands are normally less than 200 m (650 ft) deep, some hosting reefs composed of algae, sponges, corals or other sedentary organisms that slow ocean waves and create lagoons. In tropical areas with coral reefs, algae-grazing parrotfish reduce stony corals to sand, which is moved by storms, waves and tides onto beaches. These sandy beaches and their dunes then become refuges for burrowing crabs and insects, or the literal stomping grounds of shorebirds.

Tide pools reflecting sunset sky, Manzanita Beach, Manzanita, Oregon, USA

As ocean waters warm and sea levels rise in the upcoming century or beyond to make new shores, we must ask ourselves how best to preserve life on and near these places.

03

The substrates of shores today vary from muddy to sandy to rocky, with these media supporting ecological communities appropriate for their latitudes. For example, mangroves grow best in muddy and sandy sediments along tropical coasts, supplying shelter and food for shallow-marine crustaceans and fish below and for insects and birds above. In contrast, marshes are more typical of temperate coasts, their grasses firmly rooted in mud and supporting snails, crabs, wading birds and more.

Rocky shores may host remarkably diverse tide-pool communities in temperate and tropical climates, and coastal cliffs are safe places for nesting seabirds. Sandy beaches in semitropical to tropical parts of the world are where sea turtles venture onto land to dig nests and lay eggs, enabling future generations to do the same. Some of these same sandy beaches host other nesting animals, such as shorebirds on and behind dunes and horseshoe crabs in the surf. Within other coastal environments are anemones, worms, clams, snails, crustaceans, sand dollars, urchins and sea stars. Shores of all kinds also receive the flotsam and jetsam of both land and sea, with seashells and jellyfish alongside driftwood, the latter engraved by marine clams and other invertebrates adapted to this land-born material.

In the past few centuries, sudden and massive human development has heavily modified or otherwise shaped ocean shores, altering them to meet our needs or wants. Shoreline ecosystems and their biodiversity have paid a heavy price, with some vanishing altogether. Thus within our expansive view of ocean shores, we should also heed warnings of how our activities have consequences that manifest as tsunamis of irreparable change. As ocean waters warm and sea levels rise in the upcoming century or beyond to make new shores, we must ask ourselves how best to preserve life on and near these places, so that their services and wonders remain a part of our shared futures.

Rocky Shores

Some shores have little sand and are instead comprised mostly of gravel, cobbles or boulders, with their parent rocks often looming as imposing vertical cliffs. These beaches have distinctive soundscapes, from the clattering of pebbles moved by waves to the wallops of water flowing between boulders. Moreover, these rocky places offer unique conditions for ecological relationships, selecting for life hardy enough to deal with its hard surfaces.

Cobblestone beaches are far less malleable than sandy beaches; the sheer mass of the cobbles' particles means they are less likely to be moved far by waves and tides or rearranged by the activities of plants or animals. Such beaches are also less likely to hold life encrusted on or between cobble surfaces, with constant motion and crushing contacts making any chance of colonizing almost impossible. Still, these rounded rocks may hold evidence of marine lives from long-past oceans, such as those found on the Jurassic Coast near Lyme Regis in the UK, famous for its stones that reveal sections of beautifully coiled ammonite shells or isolated bones of ichthyosaurs.

Rocky shores are not necessarily composed of cliffs and cobbles, but may bear craggy edifices, with corners and edges carved by waves, tides, algae and intertidal animals. For instance, the relatively modest and low-lying limestone coasts of the Bahamas and Bermuda, which were originally built upwards by coalescing and cemented dunes, are assaulted by energetic waves. Algae specially adapted to each of these elevations impart colour-coded bands within the intertidal zone, with a lowest-lying yellow-brown strip succeeded by a black one, which in turn is topped by a grey-white band with the least amounts of algae.

Weathering and erosion of these areas is also caused by the animals that cling tightly to rocky surfaces, best represented by molluscs such as periwinkles, nerites and chitons. These molluscs feed on algae growing on rocks by scraping them off using a special tongue-like anatomical structure called a radula. This results in collateral damage to the underlying rock, with rock bits ingested as geological roughage later excreted as mud. This daily attrition by bioerosion is significant,

01

02

01 – Erratic boulders
Isle of Eigg, Inner Hebrides, Scotland, UK

02 – Tide pool and full moon
Devil's Punchbowl, Devil's Punchbowl State Natural Area, Oregon, USA

03 – Ocean waves on black-pebble beach

04 – Rocky shore
Basque Country, Spain

03

04

reducing rocky profiles while also imparting miniature valleys separated by pointed peaks and knife-edged ridges. Other animals that modify rocky intertidal zones include barnacles (see pp. 90–91), which cement themselves onto surfaces, or rock-drilling clams, which bore into the hard substrate for permanent homes, making holes that persist long after their residents' lives.

Vertical cliffs next to shores are not necessarily devoid of life either, with algae, fungi, mosses and even a few flowering plants or bushes residing on them. Animals that find safety in sea cliffs include seabirds, which nest on ledges or other horizontal places there, with elevated and hard-to-access spots effectively preventing egg and chick predators from reaching them. However, sea cliffs also become sites of tough parenting and familial loss, as hatchlings sometimes fall out of their high nests or fledglings try to fly just a few days too soon. Whenever this happens, these baby birds inadvertently supply marine carnivores with easy manna-from-heaven deliveries.

Wave Power

Waves are not only what we see on shores, but also what we hear and feel. The white-capped cresting of an incoming wave is accompanied by the sound of the rolling collapse of water hitting a sandy or rocky bottom. Depending on the speed and height of waves, this sonic heft may be a subdued splash or a booming crash followed by a hissing of water finding its way back to the ocean. When waves crest and fall onto the shore, huge volumes of water displace air while causing minor tremors under our feet. Sometimes we stand back to admire, other times we pause in the shallows, allowing the water to fall and flow around us. The effect is mesmerizing, tugging at a primal place in our consciousness. For those who take to the surf on boards, the energy of the waves is harnessed for sport.

For water to change from stillness to a wave, several factors must affect it, and nearly all are related to wind moving above and along water surfaces. Ocean waves are created by friction between air and water surfaces, with size and speed affected by wind speed, wind direction and fetch, which is the distance the wave travels across an ocean surface. Faster wind speeds create more force, piling up water that rolls down and up in circular movements, ultimately making cylinders that spin, their momentum interrupted only as they hit shallow bottoms near and on the shore.

Waves are great sculptors of shores. When waves fall on a sandy shore at an angle, their swashes pick up and push sand along the shoreline in semi-circular progressions. This taking of sand and adding it further down shore is appropriately called longshore drift. On barrier islands where waves usually come from the northeast, longshore drift thins an island's northeastern corner while fattening its southeastern corner, causing a 'drumstick' island shape. Wave impacts also help to shape rocky coasts from sheer stress, as well as by whatever might be included in a wave, which varies from sand to much larger particles. This wave-caused erosion may then work in tandem with millions of small animals, such as marine snails that cling to and scrape algae growing on the rock.

The effects of atmospheric disturbances, such as tropical hurricanes, cyclones, nor'easters or other storms, can create unusually large ocean waves that are sudden and horrific, rapidly altering coastal ecosystems and flooding coastal communities. Rarer but no less significant are tsunamis, which are caused by a combination of plate tectonics and gravity. These waves form with the rapid release of tensional energy producing an earthquake, sending invisible seismic waves from offshore epicentres that push massive volumes of water far above normal ocean surfaces, followed by these masses dropping and radiating away from the epicentre. These enormous and speedy waves – sometimes more than 30 m (100 ft) tall and exceeding 1,000 km/h (600 mph) – can temporarily emerge and submerge previously idyllic shores, wreaking destruction and ending hundreds of thousands of shore dwellers' lives, human and otherwise. Enormously destructive tsunamis were suffered by the people of Indonesia and other Indian Ocean coasts in December 2004, and by Japan's Pacific coast in March 2011. Such extremes remind us that ocean waves breaking on land can range from the soothing to the dreadful.

Deep blue tube wave

Tides

Ever since the Earth birthed its oceans, a little less than 4 billion years ago, and continents rose above the ocean surfaces, celestial bodies have worked together to make oceans regularly move onto and recede from the edges of the shore. Tides are also waves, but with much longer wavelengths – hence they are much quieter than those rendered by winds or earthquakes. Tidal movements are caused by gravitational attraction, in which matter seeking other matter while in motion exerts a pulling effect. In this instance, the greatest source of gravitational force is otherworldly, caused by Earth's nearest neighbour: its moon. Earth and moon are bound to one another as mutually attracted partners that physically redistribute enormous amounts of water on the Earth every day. Although the moon's mass is only 1.2 per cent that of the Earth, it is immense when compared to that of moons orbiting other

Example Tides Chart

Tide charts predict the high and low tides at a particular location, throughout the day, week and month. High and low tides typically occur twice a day, roughly 12 hours apart.

This example chart shows high and low tides during an approximate six-week period, with the corresponding moon phases depicted around the edge. The dots plot each day's high and low tides.

Mean high:	0.3 m (1 ft)
Mean low:	2.3 m (7½ ft)
Highest tide:	2.8 m (9 ft)
Lowest tide:	–0.6 m (–2 ft)

- ● PM Shift
- ● AM Shift

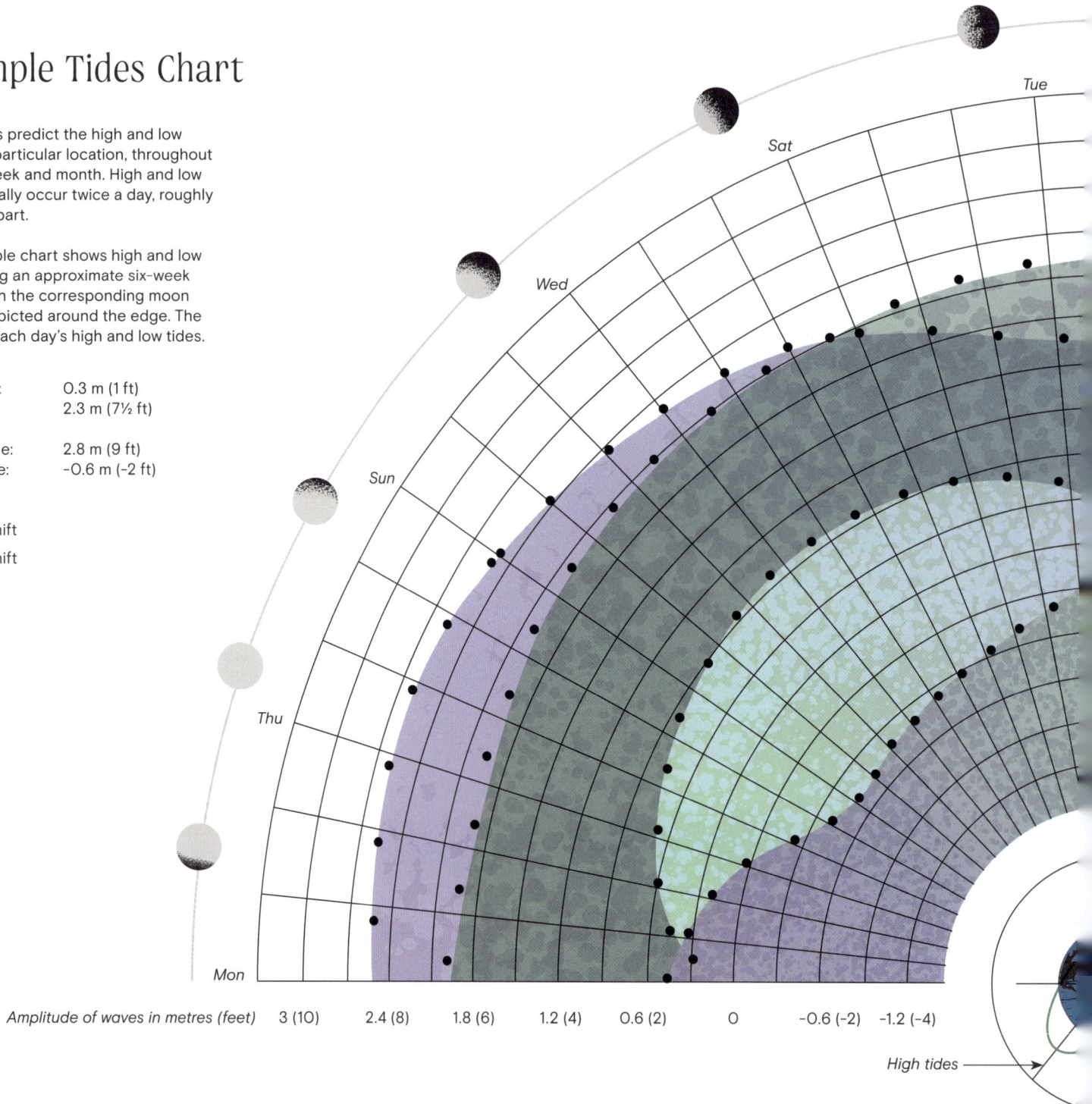

Amplitude of waves in metres (feet)   3 (10)   2.4 (8)   1.8 (6)   1.2 (4)   0.6 (2)   0   –0.6 (–2)   –1.2 (–4)

High tides ⟶

planets. Part of the moon's effect derives from its size and closeness to Earth and its oceans.

A co-conspirator with the moon in this mass movement of seawater is the sun, which, despite its greater distance from Earth, adds to this tug-and-pull of matter to matter. These combined forces are more evident during so-called spring tides – which occur approximately twice a month, not only in spring. These are extra-high tides in

which the Earth, moon and sun align to gently draw ocean waters up during the Earth's rotation, a subtle heightening that then lowers after that part of the Earth has rotated past the moon–sun alignment. Conversely, when the sun is positioned at right angles to the moon and Earth, then the gravitational forces are less, resulting in extra-low, or 'neap', tides.

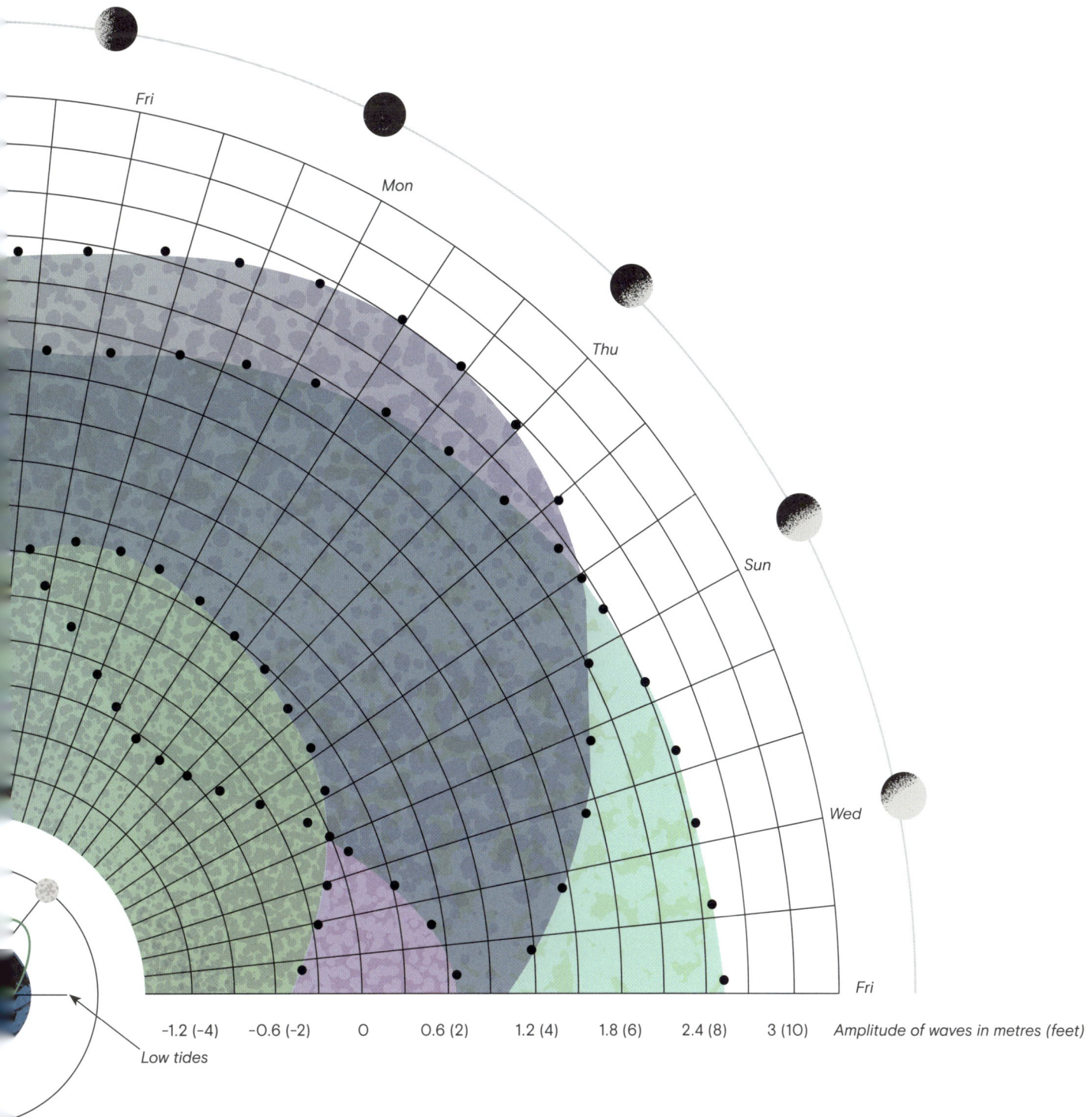

Fri

Mon

Thu

Sun

Wed

Fri

| -1.2 (-4) | -0.6 (-2) | 0 | 0.6 (2) | 1.2 (4) | 1.8 (6) | 2.4 (8) | 3 (10) | *Amplitude of waves in metres (feet)* |

Low tides

Gifts from Sea to Land

Waves and tides sometimes make special deliveries to the land from the sea. The items they deposit may have once grown and lived on these same lands, making their return a sort of ecological regifting. Pieces of former forests that float in the ocean before coming back ashore are collectively termed driftwood. Ranging from bits of crumbled wood looking like discarded coffee grounds, to nearly intact trees with roots and all, driftwood reminds us of how the land donates parts of itself to the ocean before being returned, while also bearing evidence of life histories etched into its surfaces.

Before taking its wayward voyages on the open ocean, driftwood commonly originates as dead trees near rivers that empty their contents into the sea. These trees were far from barren places – their interiors and exteriors bear evidence that they hosted thousands of lives, such as borings made by wood-eating insects and insect-eating birds. Woodpeckers will have treated some of these trees as nurseries, in which parents carved out cavities used for nesting and raising their young.

A seafaring tree may also take on marine passengers, such as barnacles attached to outer surfaces, small wood-eating crustaceans called gribbles and wood-drilling clams (see p. 86). Coming full circle, decomposed driftwood reintroduces nutrients to food webs on or near shores, thus affecting plants and animals living in those places.

01 – Small barnacles and
lacy bryozoans on wood
Black Sea beach

02 – Barnacles on driftwood
washed ashore
Malaysia

Intertidal Zones

Tides along different shores are classified by vertical differences in low and high tides, with three categories: microtidal (less than 2 m / 6½ ft difference); mesotidal (2–4 m / 6½–13 ft); and macrotidal (tidal ranges exceeding 4 m / 13 ft). Such vertical differences then translate to intertidal zones – alternately exposed and inundated areas of mud, sand or rock bracketed by low and high tides. Most shores are microtidal, but mesotidal and macrotidal regimes have more pronounced effects, with all shaped by regular flooding and ebbing augmented by wind-driven waves. Microtidal ranges play a significant role in the formation and maintenance of nearly all barrier islands, from the Arctic Circle to the tropics. For those barrier islands affected by mesotidal effects, such as along the coast of Georgia, USA, tides nourish and sustain broad, grassy plains of salt marshes between the islands and mainland shores. Macrotidal environments are rare but spectacular, draining intertidal beaches or mudflats over vast areas, only to return with voluminous amounts of water and momentum. Places with macrotidal regimes include the Inchon tidal flats in Korea, the Bay of Fundy in Nova Scotia, Canada, and parts of Western Australia.

Life along shores in and above their intertidal ranges is well adapted to these daily and monthly fluctuations, whether exposed to air or completely submerged. For example, many marine-dwelling animals that live in rocky intertidal zones – such as anemones, crabs, sea stars and urchins – find refuge in hollowed-out spaces that retain enough water for their denizens to survive until the next flood tide. Along sandy or muddy shores, clams, snails, crustaceans and sea cucumbers may burrow into moist sediments below, neatly avoiding the conjoined perils of dehydration and predation as they wait for the next cycle. However, not all are safe from animals that venture onto emergent surfaces for all-you-can-eat seafood feasts during low tides. Shorebirds are among the most rapacious hunters after a high tide exposes broad areas and their potential bounty, but they are sometimes rivalled by other animals adapted for semi-terrestrial lifestyles, such as crabs. Tides also work with wind-blown waves to deliver floating passengers to land, whether alive or dead, and ranging from floating *Sargassum* seaweeds to once-majestic whales. Tides thus give us glimpses of what lies far beyond the shore while beguiling us as ever-shifting stages for ecological dramas.

Clachtoll Beach, Scotland, UK

Fiddler crabs scavenging at low tide, Florida, USA

Crab feeding pellets and trails, Australia

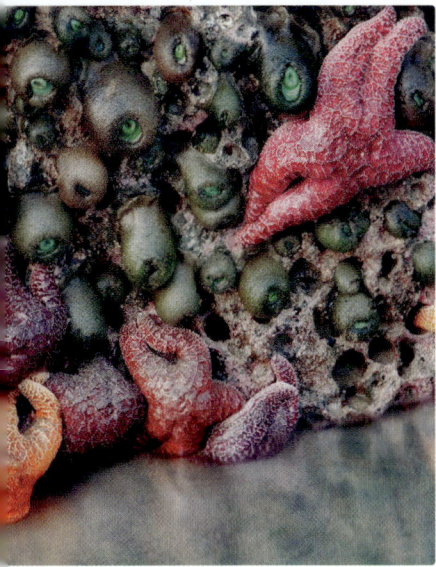
Sea stars and sea anemones at low tide,
Oregon, USA

Aerial view of the Wadden Sea at low tide, Germany

Rough ocean with waves crashing against cliffs, Ireland

Aerial view of winter tidal flats, South Korea

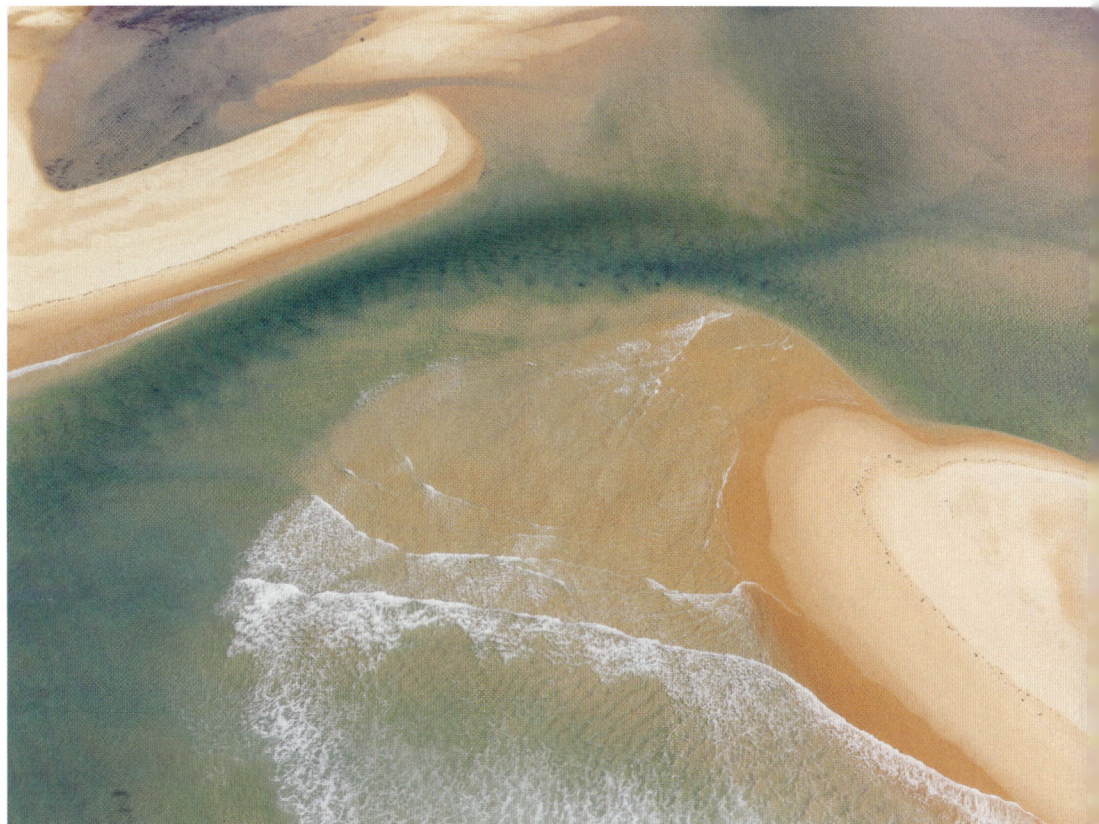
Aerial view of sandy beaches and tidal water patterns, Australia

Shifting Sands

Sand is nearly synonymous with seashores, especially beaches. The composition of sand is often assumed to be quartz, a durable silicate mineral. Yet not all beaches are sandy; nor is sand restricted to shores or exclusively composed of quartz. Sand is a size term, which sedimentologists define as sedimentary particles between 0.06 and 2 mm (1/500–1/16 in) wide; further subdivisions of this size range are given modifiers such as very fine sand, medium sand and very coarse sand. Sediments smaller than sand are classified as silt- or clay-sized, whereas sediments larger than sand are called gravel, cobbles or boulders.

Sand can be further described by the outer surfaces of individual grains and their shapes. Some sand grains, known as rounded, have smooth exteriors lacking corners; others, known as angular, have points. Grains with smoothed points are termed subangular or subrounded. This continuum of sand textures is also accompanied by shapes compared to an idealized sphere; hence sand grains can have high or low sphericity. When scooping a handful of sand to examine it, a closer look may reveal its sizes, shapes, colours and other variations that depend on their sedimentary journeys.

Coastal sand compositions are also differentiated by the way they reflect their sources, varying from distant mountains to rocky shores to offshore reefs. For those shores in which sands are primarily quartz, they may represent multiple movements of these minerals removed from silicate-rich rocks far up rivers, or they can be more locally derived, coming from nearby rocks or ancient sandy deposits. In Brittany, France, for example, the Côte de Granit Rose (pink granite coast) boasts striking pink sands, their distinctive colour derived from local rocks composed of quartz and potassium feldspars. Darker grains within mostly quartz sands are often composed of denser iron- and titanium-bearing minerals, leading to their concentrations where less dense quartz sand was blown away by winds. In places with volcanoes, beach sands can become more exotic, with dark green hues bestowed by olivine.

As for many tropical beaches, whitish calcium-carbonate sand is mostly made by offshore life, whether precipitated by algae or invertebrates, or broken down by rock-eating animals, such as parrotfish (see p. 123), which munch on reefs and excrete carbonate sand. A special kind of sand formed in tropical areas consists of ooids – minuscule, near-spherical grains of calcium carbonate that accrete in thin layers around shell bits as warm-water waves move them back and forth. Modern ooid-laden sands make up gorgeous underwater dunes in the shallows of the Bahamas and the Persian Gulf, but ooid sands were also produced by warm-water shallow seas more than 300 million years ago.

For barrier islands and some other shoreline environments, sand gives shape to their bodies, often with rivers, waves and tides feeding these sediments to them. Yet much like our own bodies, denying these shores their sandy sustenance leads to thinning and eventual starvation. This effect is most dramatically demonstrated when upstream dams halt the flow of sand downriver, causing formerly robust barrier islands to thin or disappear entirely. Whether shore sands come from mountains and rivers or from nearby carbonate-mineral sources, cutting their supply will change those environments into something different.

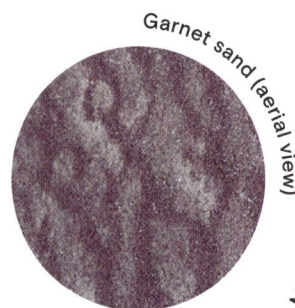

Garnet sand (aerial view)

Mineral sand (aerial view)

Mineral sand (magnified)

Quartz sand grains (magnified)

Olivine sand (magnified)

Glass beach sand (slightly magnified)

Garnetiferous sand (slightly magnified)

Gypsum sand (aerial view)

Sand fragments from old pieces of glass bottle (magnified)

Mangroves

Mangrove forests or swamps serve similar functions to marshes (see pp. 40–41) but differ in specific ways. Their main distinctions are in their plant communities, with mangrove forests dominated by, well, mangroves, which are saltwater-adapted flowering shrubs and trees. Mangroves also require narrower temperature ranges for their growth, with most restricted to subtropical and tropical coasts throughout the world. Nonetheless, as in salt marshes, the robust and complicated networks of mangrove root systems help to retain fine-grained sediments while baffling sediment-laden currents. This slowing of flowing water allows for more mud deposition, which along with burrowing crustaceans helps to churn organics into the sediment and allow oxygenated waters to reach deep down into the mud. Upper parts of mangrove root systems, consisting of buttresses and pneumatophores – vertical roots coming up from the ground that serve as 'breathing tubes' for mangroves – also rise above the sediment surface and breach high-tide marks, forming habitats that effectively hide small crabs and fish while barring entry to larger predators.

Red mangrove tree
Rhizophora mangle

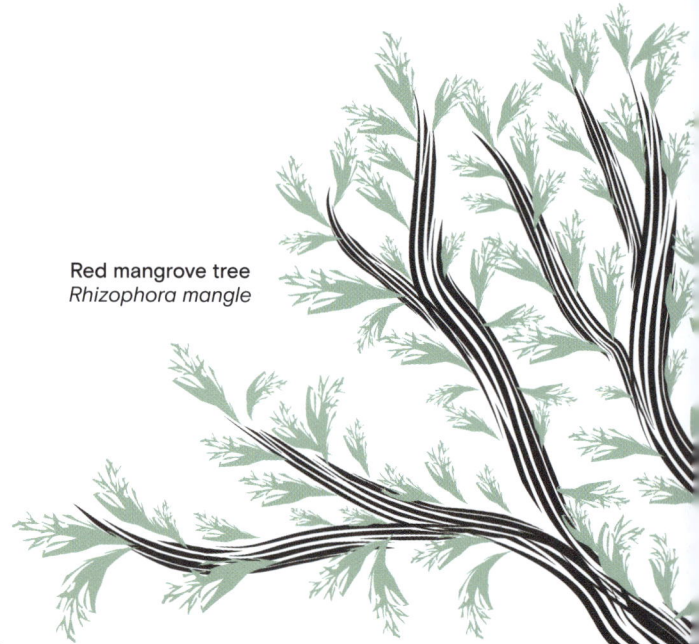

Brown pelican
Pelecanus occidentalis

American crocodile
Crocodylus acutus

Detritus

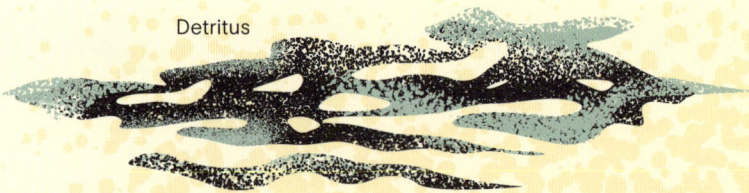

Pink shrimp
Farfantepenaeus duorarum

Great egret
Ardea alba

Little red bat
Lasiurus minor

Great blue heron
Ardea herodias

Mangrove forests are already valuable coastal ecosystems for their high biodiversity, but much like marshes, they also directly aid humans on both local and global scales. For one, these forests slow down storm waves, helping to protect coastal communities on their landward sides. More broadly, mangrove forests are among the most efficient and effective ecosystems for capturing and holding on to carbon, hence helping combat climate change. Given their beauty and benefits, mangrove forests deserve our appreciation and protection.

Grey snapper
Lutjanus griseus

Bocourt swimming crab
Callinectes bocourti

Hard clam
Mercenaria mercenaria

An Ecosystem of Mud

You might assume that shores composed of mud would be washed away with each tidal exchange or by coastal waves, their suspended minerals and organics imparting various coffee and chocolate hues to local waters. Yet much of this mud is retained, thanks to plant life holding it down and animals producing packets of mud. These muddy shoreline environments are best represented by marshes and mangrove forests (see pp. 38–39).

Coastal marshes are more properly called salt marshes in recognition of their saline surroundings and grass-dominated plant communities. However, only a few species of grasses are specially adapted to twice-daily baths by high tides and sun exposure by low tides. In this sense, they are like perennially wet-and-dry prairies and, like land-bound prairies, they support their own distinctive faunas. Salt marshes along the eastern coast of the USA stretch from Florida to Maine on mainland shorelines, as well as on mainland sides of barrier islands. Their grasses, such as *Spartina*, have complex and extensive root systems that anchor mud and resist its erosion. In these ecosystems, periwinkle snails graze on algae growing on grass leaves and stalks, tearing off tiny leafy bits that fall into the mud and water below. These and other organics in turn are ingested by mussels, oysters and other filter-feeding animals attached to submerged bottoms, where they also pump out significant amounts of neatly packaged sand-sized mud, held together by mucus. The organic rain from above also feeds fiddler crabs, which scrape algal films off marsh surfaces during low tides; their numerous burrows aerate mud and allow spaces for other small crustaceans and marine worms. Marshes are also cut by meandering tidal creeks that look and behave much like miniature rivers, often laterally migrating their oyster-laden banks into twisting bends.

Along with their important role in maintaining coastal biodiversity and their ecological functions, salt marshes help humans, too. For example, marshes slow and otherwise absorb the energy of destructive waves, such as those produced by hurricanes and other storms. They also serve as nurseries for many shallow-marine animals, including shrimp, crabs and fish. Even visitors to marshes benefit from them, with small songbirds weaving nests within and alongside them, and raccoons, wading birds and dolphins preying on their animal abundance.

The ecosystem services and great beauty offered by marshes and mangrove forests demonstrate the intrinsic value of these environments. However, they are both endangered worldwide by coastal development, industry, pollution and other human activities, and thus require focused attention on their preservation so future generations of human and nonhuman lives can continue to benefit from them in perpetuity.

01

01 – Salt-marsh landscape

02 – Aerial view of marsh wetland abstraction of salt and seawater
Rachel Carson Wildlife Sanctuary, Wells, Maine, USA

03 – Salt marshes at dusk

Secrets of Dunes

Most shores with sandy beaches have upslope neighbours called dunes. Although coastal dunes are not in the sea, they often exist because of the ocean, with their sand supplied by winds, waves, tides and storms. At least some dunes are also affected by nearby marine life that walks across or digs into them, such as crabs and sea turtles. Coastal dune interiors can also host land plants and animals – from grasses to insects to mice to nesting shorebirds – while providing elevated surfaces above beaches for land animals to tread. Because of their height above beaches, dunes can serve as barriers to waves and tides, excluding or at least slowing the passage of ocean waters into landward environments behind the dunes. In some cases, though, these defences are breached by storm surges that cut through and flatten what was once there, allowing for the genesis of new dunes a bit further inland.

The anatomy of a typical coastal dune is surprisingly complex, especially when infused with life. Dune outer surfaces form hillocks that often coalesce, creating undulating linear ridges that are more-or-less parallel to shore. Wherever a dune ridge is eroded by the shore, exposed interiors reveal histories of accretion and erosion, with thin curved or straight lines marking surfaces where wind-blown sand tumbled up and over dune crests or scoured lower areas that were eventually filled by more layers of sand. These horizontal and angled surfaces collectively make cross-bedding, a type of sedimentary structure also found in inland dunes, along rivers or in subtidal environments. While looking at an exposed dune interior, you might also see where its cross-bedding was interrupted by roots or the burrows of insects, ghost crabs and mice. If you are very lucky, you may see the eroded section of a sea-turtle hole nest with empty eggshells abandoned by hatchlings, testifying to a season when these turtle tykes left their surrogate-mother dunes and journeyed from the land to the sea.

Because coastal dunes lie between oceanic and land environments, they are necessarily born from processes supplied by both. Clastic sand – sand particles coming from rocks composed of quartz and other silicate minerals – is brought to shores by rivers that empty into the shallow sea to form deltas, estuaries or barrier islands. In more subtropical and tropical environments, sand in dunes is mostly composed of calcium carbonate minerals, such as aragonite and calcite, that are produced just offshore by ooids, algae, corals, molluscs, sea urchins and parrotfish. Barrier islands commonly form in places with sufficient sand to accumulate into dune ridges parallel with the shore; these ridges are then isolated as seawater fills the low depressions behind the dunes, forming a lagoon between them and the mainland.

Once barrier islands are in place, their dunes often maintain them even as storms, waves and tides do their best to erode them. Wherever dune-loving plants are available, such as sea oats and creeping vines, these help to hold down dune sands and stabilize them. In contrast, dunes with little or no vegetation are constantly reshaping. Unvegetated and topmost dune surfaces also might be adorned with wind ripples, asymmetrical waveforms with crests parallel to one another. Interrupting the geometric repetition of wind ripples are the tracks, trails and shallow burrows of animals that somehow make a living in these thinly confined ribbons of sand.

Aerial view of Skeleton Coast sand dunes meeting the waves of the Atlantic Ocean
Skeleton Coast, Namibia

Welcome to the Nest

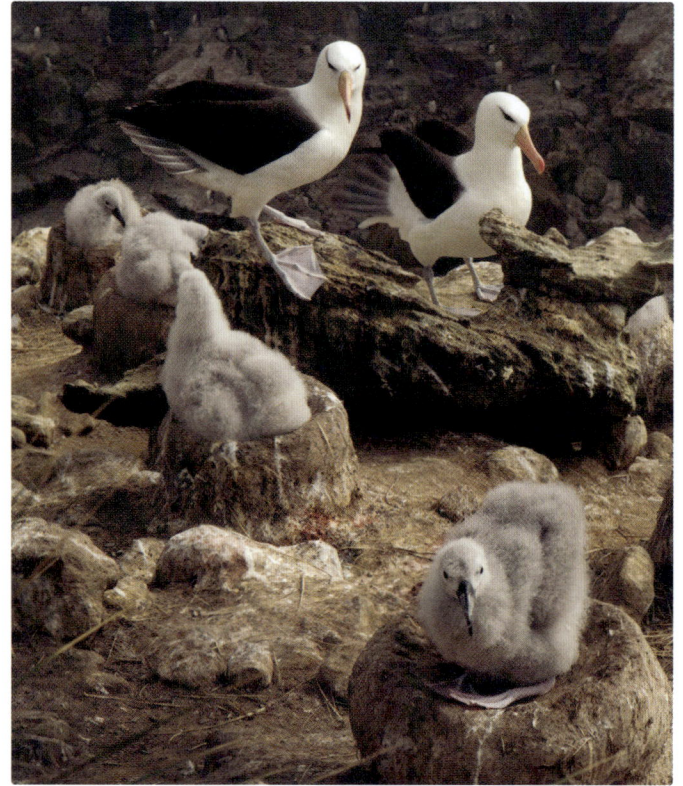

Adult black-browed albatrosses with chicks in nest, Falkland Islands

Many animals that spend most of their lives in the ocean use shores as their nurseries. Perhaps the most famous of these creatures are sea turtles. Mother sea turtles all began their lives by hatching and kicking their way out of a sand-buried nest, emerging, and somehow surviving the perilous and well-documented trek from onshore nest to the sea. Some of these survivors then found enough food to make it through more than thirty years of swimming in the open ocean, before successfully mating and often returning to the same beach where their lives began to lay their own eggs, echoing sea-turtle behaviours from the past 100 million years. When the turtles gather in the thousands, as they do on the Pacific Ocean beaches of Central America, it is a truly remarkable sight, a glorious event of mass resilience and maternal relief.

Despite the longer evolutionary journeys of sea turtles, the most widespread and varied of vertebrates that nest on shores worldwide are shorebirds. Gulls, cormorants, pelicans, terns, black skimmers, oystercatchers, plovers, puffins, shearwaters, albatrosses and penguins are all examples of birds that make nests for protecting and incubating eggs on coasts. Unlike sea turtles, though, shorebirds stay with their eggs and chicks, and often work as male–female pairs. Their nests are as diverse as their makers, ranging from the simple scraped depressions of oystercatchers and plovers in dune sands to the metre-long burrows of puffins, shearwaters and fairy penguins. And in some instances, mother seabirds are among the oldest-known mothers, most famously represented by a Laysan albatross nicknamed 'Wisdom'. Wisdom was about five years old when tagged in 1956, and was spotted incubating an egg (presumably hers) in 2020, making her at least seventy years old.

Gentoo penguins nest in guano and stones in a rookery, Neko Harbour, Antarctica

Variable oystercatchers with their nest

Iceland gulls resting on nests on a rock by the ocean

Atlantic puffins at Drumhollistan, Scotland, UK

Hatchlings of the olive ridley sea turtle emerging from a nest in Costa Rica

THE SHORE

Horseshoe Crabs from the Far Past

Given both their appearance and their behaviours, limulids – also called horseshoe crabs – are perhaps the most primeval of animals found on Atlantic or Pacific shores. Horseshoe crabs are not true crabs, nor are they even crustaceans, but are distantly related to modern land-dwelling spiders and scorpions. Their ancestry affirms their ancient heritage, with limulid body parts and their distinctive tracks preserved in the fossil record more than 400 million years ago, during the Palaeozoic era. Nevertheless, only four species made it to today for us to appreciate, with three in Southeast Asia and one (*Limulus polyphemus*) along the east coast of North America.

These large, brown to forest-green invertebrates have a tough but flexible organic exterior, with a prominent head shield (prosoma) bearing a pair of compound eyes on left and right sides, as well as smaller light-sensitive organs between these and on the underside of their bodies. Behind the prosoma is a smaller trapezoidal part (opisthosoma) that hinges with the head, and behind that part is a long spiky tail (telson). Of their dozen legs, five pairs are for walking and the forward-most pair (chelicerae) is for eating. The front legs pick up potential food items on the ocean floor, such as small worms, clams and crustaceans, and pass these morsels to the mouth, located between the walking legs. But instead of simply shoving these items into its maw, the horseshoe crab uses stiff bristles on its legs to shred the food as a sort of chewing before swallowing.

As for horseshoe-crab behaviours, they are the only marine invertebrates that must reproduce on land. Hence, they are adaptable enough to leave behind oceanic support and walk on beaches, leaving their tracks alongside those of sea turtles and shorebirds. Female horseshoe crabs, which are larger than the males, crawl into sandy intertidal areas, where they lay tens of thousands of coarse-sand-sized, turquoise eggs. Meanwhile, smaller males move around, near and onto females, hitching a ride by grasping the female's opisthosoma with their front legs and fitting the front edge of their head into a groove on her rear. If one of the males succeeds in making such a connection and holds on long enough to be in the right place when the female lays her eggs in the sand, then he releases sperm to fertilize the eggs. Successful fertilizations result in tiny baby horseshoe crabs that live most of their lives in shallow water offshore or in intertidal areas. The mass spawnings of horseshoe crabs are spectacular to behold on early summertime beaches in the mid-Atlantic states of the USA, such as Virginia, Maryland and Delaware, but they are also commonly encountered in their full species range.

The seasonality of these horseshoe-crab emergences has another effect on beaches, in that their copious eggs provide fuel for animals on epic journeys. In North America, the consumers of these eggs are shorebirds, mostly red knots and ruddy turnstones, which time their long-distance migrations with limulid spawning. We do not know when such linkages between migrating shorebirds and mating horseshoe crabs first happened, but we can reasonably speculate that they reach back more than 100 million years ago to the Mesozoic era, in which new avian dinosaurs ate their fill of Palaeozoic-derived eggs.

Horseshoe crab
Limulus polyphemus
A Gombessa Expedition, led by Laurent Ballesta. In the Philippines, Pangatalan Island and its totem animal, the horseshoe crab, symbolize survival and ecological restoration.

Traces

Every shore holds traces of what lives there. These range from glaringly obvious, such as human footprints, to subtle, for example the shallow raspings of a rock surface made by algae-eating snails. But these imprints all tell stories of animals' daily lives, entire lifetimes or afterlives. Shore traces can be easily classified by common descriptors of their making. Among these are: burrows (excavations of sediment); borings, created by the breaking down of solid rock, shells or wood; trails (pathways formed on sediment surfaces by legless animals); and tracks, made by animals with legs. On some shores, all four of these may be present; on others, one or two may dominate.

For some burrowing animals, such as ghost shrimp, a burrow top on a sedimentary surface is a mere hint of what lies beneath. On a sandy beach, a typical ghost shrimp burrow looks like a tiny shield volcano composed of beach sand with a pencil-thin hole in its centre, its flanks adorned by dark mud pellets. And where there is one burrow mound, there are many, with perhaps hundreds within a square metre visible at low tide. The burrow shaft may continue down into the sand for several metres until it connects with a multilevel network of tunnels. Multiply these by the thousands, and you soon realize your seemingly deserted beach holds teeming metropolises beneath your feet.

In coastal dunes, ghost crabs are the main inhabitants, digging metre-deep shafts with rounded openings. Sometimes abandoned ghost crab burrows are usurped by dune mice, leaving petite footprints and tail imprints on the sand. Coastal dunes may also host the burrows of insects that dare to feed and breed close to salty water, their excavations mixing with those of marine-related animals.

Borings are ubiquitous in rocky shores, where rock-eroding sponges, snails, clams, chitons, barnacles and polychaete worms may live, their activities preserved as irregular pockmarked and holey surfaces. Other borings may be evident as perfectly circular, bevelled holes in clam and snail shells, telling of predatory moon snails that drilled into their living owners for a meal. Borings in driftwood may reflect multiple origins, from forest-dwelling insects to seafaring clams.

Trails are often made by marine worms or gastropods, but in some instances are left by sea stars and sand dollars stranded by a low tide. Snails of all sizes and types move via expansions and contractions of their muscular feet, and they may glide along on mucus-lined paths, with this layer of slime helping to lubricate their passage while preventing abrasion or other injury. For those snails that ingest algae off sandy, muddy or rocky surfaces, they apply their radulae (rasping tongue-like structure; see p. 24) and change rock to mud.

Tracks are left by the many-legged – such as horseshoe crabs (see pp. 46–47), hermit crabs and true crabs – and the two-legged. Their marks are distinguished by the number and shapes or points of the leg impressions they leave; true crabs, for example, have trackways with sets of four points on either side, vestiges of their side-scuttling movements. In the centre, you might see one or two drag marks made by paired claws held close to the sandy surface. Four-legged and marine-dwelling animals, such as sea turtles, otters and seals, may also leave their journeys embroidered on sandy or muddy surfaces, often joined by the paired tracks of shorebirds and people.

Common limpet adult, with feeding tracks, attached to rock at low tide

Sea-snail trails on a sandy beach

Limestone crisscrossed with snail trails at low tide

Sea driftwood with borings by 'shipworm' clams

Rough piddock clam borings in solid rock

Maculated ivory whelk

Pearl turbo

Ventral harp

Endive murex

Common tower

Banded marble cone

Spotted triton

Strawberry trochus

Articulate harp

Pearly nautilus

Paua abalone

Atlantic fig snail

Rare-spined murex

Violet clam

Green turban

Chiragra spider conch

Sea urchin shell

Peruvian scallop

Decoding Shells

Beachcombing is a popular pastime, and one of its delights is the discovery of washed-up shells. These shells may first invite our attention with their colourful patterns and forms, and after picking them up, their textures – ribbing, coils, knobs or spiky projections – also give our fingers more ways to read them. Any child knows that if you hold an empty seashell to your ear you can hear the ocean inside – or so it seems. Occasionally such shells act as hosts for newer marine lives that attach to them, ranging from algae to barnacles to sea anemones. Or their original forms may have been modified by boring sponges, worms, snails and other marine animals. Shells can thus express complex and fascinating histories to accompany their sensory appeal, so perhaps it is not surprising that they have inspired humans' artistic impulses for thousands of years by serving as jewelry and other personal adornments.

Although 'shell' is a generic term applied to a hard outer layer covering a soft interior (think of eggshells, for example), in this instance it refers to the former body parts of molluscs. With only a few exceptions, modern shells come from two evolutionarily related groups of molluscs: bivalves and gastropods. Bivalves (mostly clams) and gastropods (mostly snails) last shared a common ancestor about 500 million years ago, and once evolved, each group diversified into a stunning array of forms while also occupying a broad range of environments. All bivalves have shells, but not all gastropods do, with some swimming shell-free in oceans as nudibranchs (see pp. 120–21) or sliming along in gardens and other land environments as slugs.

Shells can also contain stories of hard lives and afterlives. Look for those that seem less than perfect or incomplete, with ragged lines telling of an escape from a crab attack, small circular holes inflicted by predatory snails, or barnacles and other creatures that latched onto their hulls long after death. Some snail shells also have polished spots on their lowermost surfaces, showing where hermit crabs lent legs to move about in completely different ways. All such clues lend to deeper insights into how these protective edifices allowed some shelled lives to persist, while others succumbed to ecological and evolutionary forces.

Fossils

Modern seashores not only host a variety of life, but also may hold evidence of past seashores or other marine environments through their fossils. Some such places are defined by sedimentary rocks exposed along the shore, such as the famous 'Ammonite Pavement' at Lyme Regis in southern England. This limestone surface and its dislodged boulders – eroded by waves and tides over the years – contain thousands of large ammonites, their exquisitely expressed spirals and chambers revealing these seafaring denizens of the Jurassic. In other coastal settings, marine fossils many be much more numerous but not so visible, like those in the White Cliffs of Dover, UK. These sheer cliffs, which have inspired poetry, songs and other creative musings over the years, are composed of trillions of the microscopic remains of coccolithophores, spherical algae made of calcium-rich plates (coccoliths) that floated in Cretaceous oceans. Significantly younger fossil shores are in southern Florida and the Bahamas, where coral reefs lived during sea-level highs of the Pleistocene epoch. The now-grey stony skeletons of corals, clams, snails and other reef dwellers testify to formerly colourful and diverse ecosystems that flourished in seas that were over 5 m (16 ft) deeper than they are today. While walking on the ledges of these reefs, one can look up and imagine sunlight filtering through a warm blanket of water above.

Shorelines composed of mere sand also may be sources of fossils that weather out of older oceanic sediments and are washed onshore. Many beaches on the southeastern coast of the United States often include fragmented and blackened fossil mollusc shells mixed in with the more complete and variegated shells of the recently dead. The blackened shells, suffused with monosulphides, are exhumed from offshore or shoreward formations of Pleistocene, Pliocene or Miocene sediments by waves, tides and storms. More durable and denser fossils of vertebrates from these times may include the flat grinding teeth of stingrays, or the bladed and serrated teeth of sharks. Perhaps the most prized of finds for people who seek these fossils are the palm-sized teeth of the greatest ocean predator of all time, *Otodus megalodon*, sometimes simply called 'megalodon' ('big tooth').

02

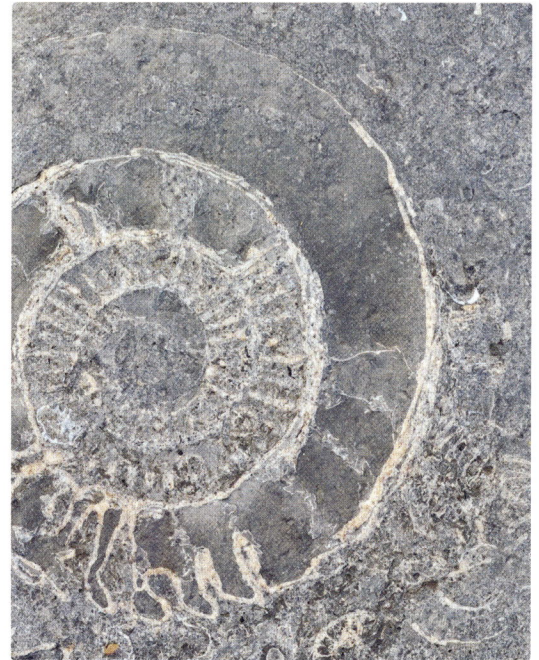

01 – Fossil coral colony
Egypt, Red Sea

02 – Large ammonite fossil embedded in rock
Pinhay Bay, near Lyme Regis, Jurassic Coast, Dorset, UK

03 – Fern plant fossils from the Bay of Fundy Basin, North America, one of the most fossil-rich deposits in the western hemisphere

How Oceans Are Made

The Earth's surface is constantly in motion. The lithosphere – the solid, cooler and outer part of our planet – is broken up into tectonic 'plates', which interact with one another in different ways. As heat from the Earth's core rises, the plates move. Convection of hotter and more plastically flowing 'asthenosphere' underneath the plates causes this movement, and is responsible for the most dramatic geological activities on Earth. From example, when tectonic plates are split apart from each other at divergent boundaries, seafloor spreading occurs – the process by which our oceans are created. The huge fissures left by these parting plates allow molten rock to rise, and as the lava cools, it creates a ridge, marking where seafloor spreading occurs. Seafloor spreading is like a conveyor belt, creating new seafloor at the ridge, which may then be destroyed later when one plate moves underneath another at ocean trenches. The Mid-Atlantic Ridge, which started about 200 million years ago and now runs down the middle of the Atlantic Ocean, is perhaps the greatest example of this remarkable phenomenon. That volcanic hotspots within a continent can become an ever-widening ocean over time is startling, one of many ways that plate tectonics has shaped our world.

The Tectonic System

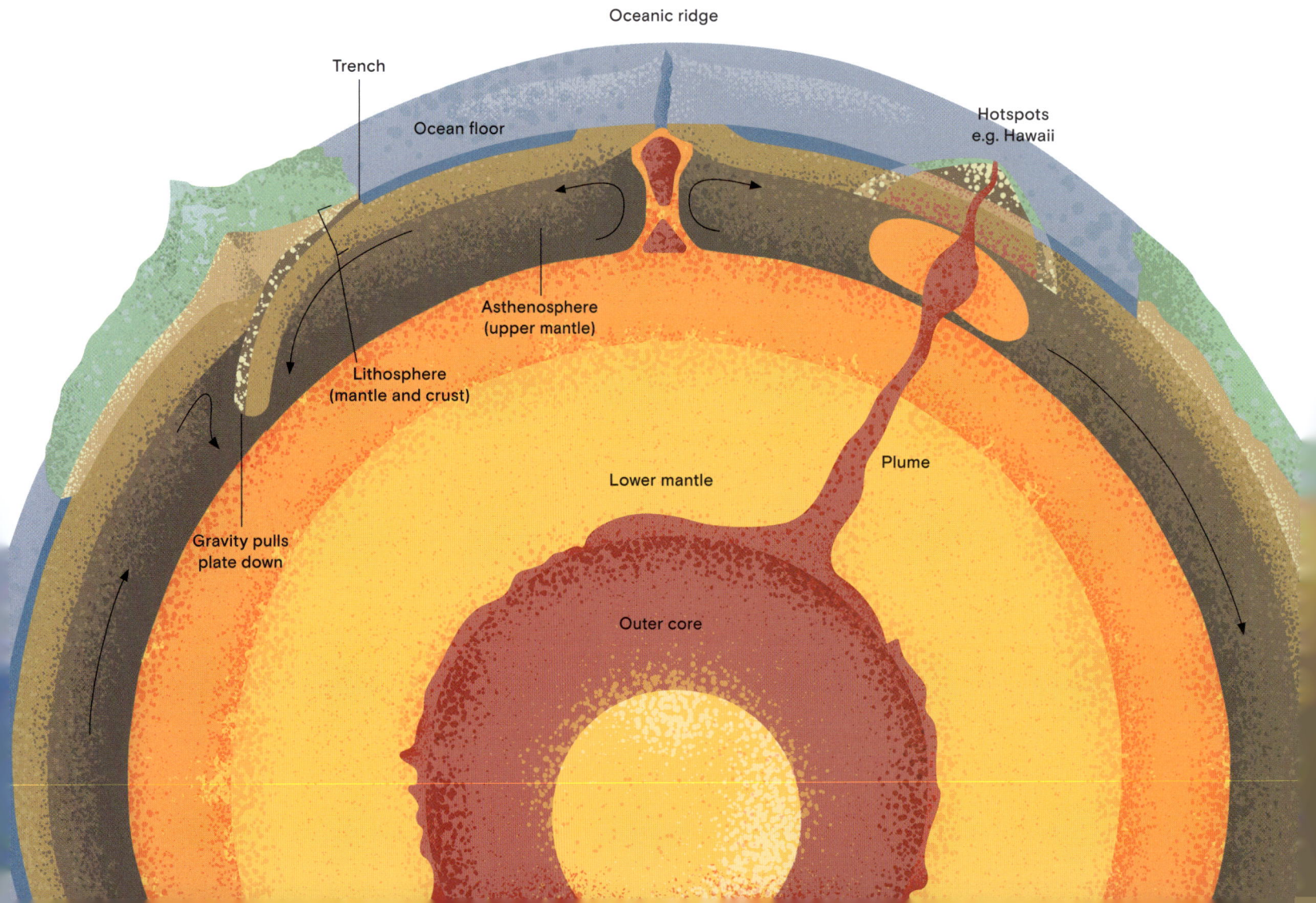

Oceanic ridge

Trench

Ocean floor

Hotspots
e.g. Hawaii

Asthenosphere
(upper mantle)

Lithosphere
(mantle and crust)

Gravity pulls
plate down

Plume

Lower mantle

Outer core

How Seafloor Spreading Occurs

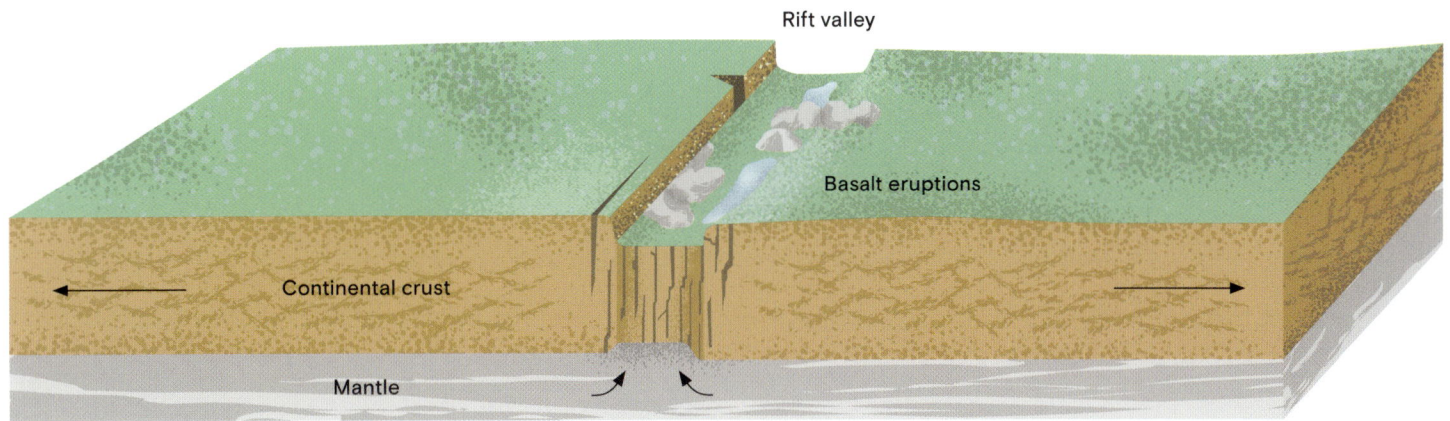

Rift valley

Basalt eruptions

Continental crust

Mantle

A fissure is created when a
continental plate is broken apart.

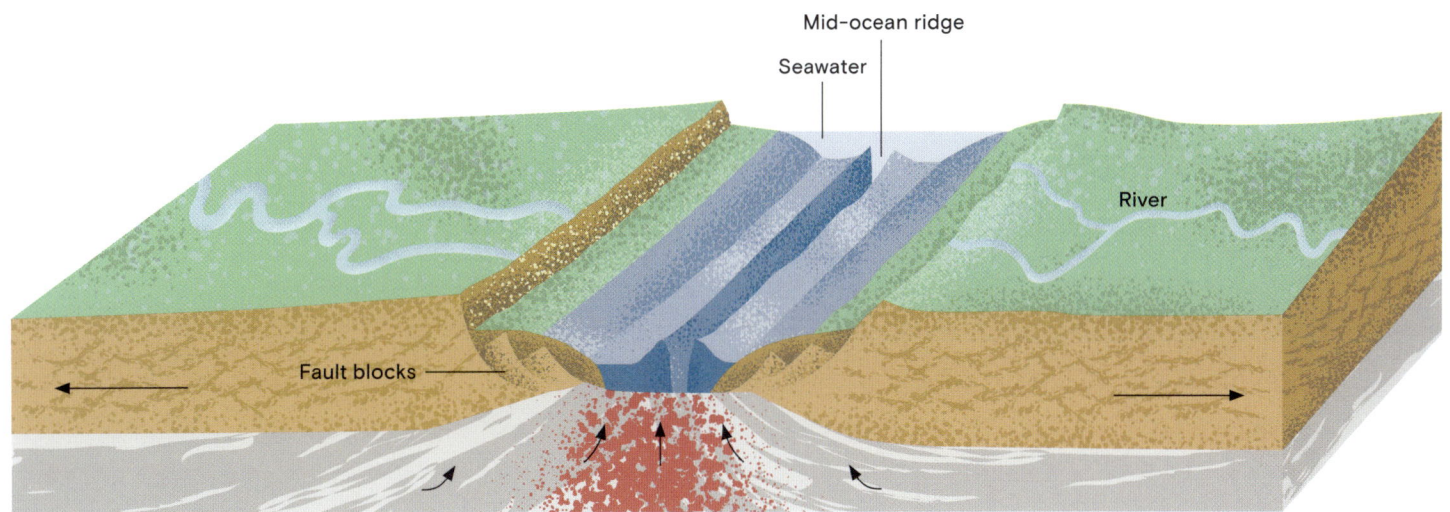

Mid-ocean ridge

Seawater

River

Fault blocks

Molten rock rises through the
fissure. When the lava cools, new
seafloor is created at the ridge.

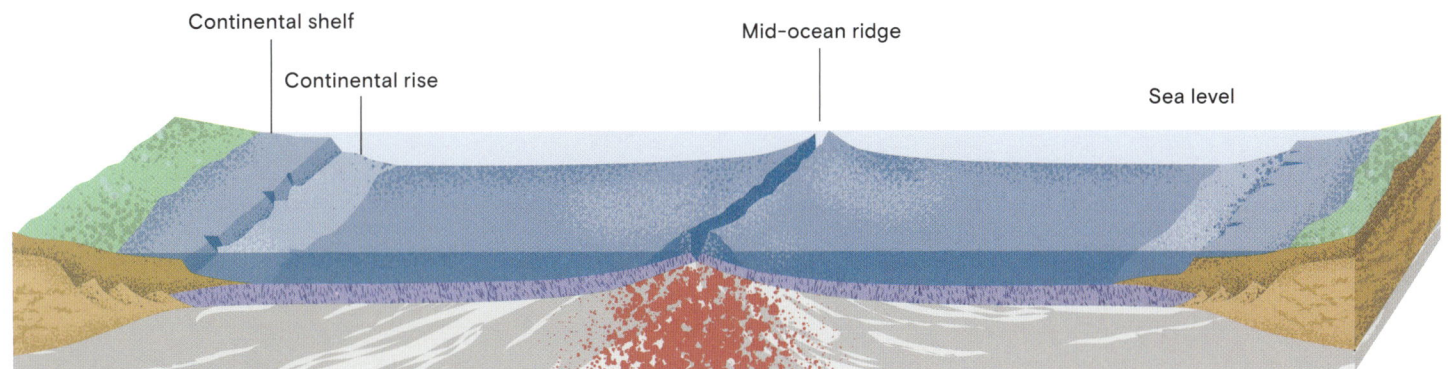

Continental shelf

Continental rise

Mid-ocean ridge

Sea level

As the process continues, the
seafloor grows as it spreads away
from either side of the ridge.

The Surface

Introduction

Life at the surface of the ocean drifts between worlds, having little say in where it is carried or the seas it inhabits. Surface creatures are adapted to thriving in the harshest atmospheric conditions, from extreme and unrelenting sun to driving winds or glass-still waters. They may be pummelled by heavy rain or cooked by heat. They may be soaked in slicks of fresh water or crusted in evaporating salt. They may drift through the sweltering tropics or be pushed to the frigid north. Waves, air bubbles and sea foam roil and tumble their bodies. Yet they remain at the surface, part of an ecosystem that has existed for hundreds of millions of years.

The surface – or neustonic zone – extends roughly 1 m (3¼ ft) above and below the waves. It serves as a nursery ground, a meeting place and a safe haven, filled with floating forests that drift with the waves. This zone has been more impacted by humans than any other region of the sea. It is the ocean's skin and lungs, the surface through which the water breathes. Yet despite its proximity to our own world, we know comparatively little about the surface. How could this be?

For decades, marine biologists have travelled through the surface into deeper water to study coral reefs or abyssal ecosystems. Yet few paused at the air–sea interface. Soviet scientists, such as A. I. Savilov or Y. P. Zaitsev, were pioneers in the field of neustonology – the study of surface life. When the Soviet Union collapsed, much of their work became inaccessible to anyone outside the region. The Soviets conducted several large-scale Pacific expeditions, while scientists from other countries were mostly based on shore. Neustonic species do sometimes wash up near coastal zones, and in some regions these shore 'spills' are associated with particular seasonal or wind patterns. But even in the best seasons and with the most favourable weather, surface life might wash up on beaches for only a few hours or days, before a subtle change in wind or tide carries the rest back out to sea. For all these reasons, the surface is, paradoxically, difficult to reach.

Despite this, what we know of the ocean's surface presents a remarkable picture of a strange and alien ecosystem. There are three types of surface life. There are permanent autonomous floating residents of the ocean's surface, or 'obligate neuston'. These include the strange seaweed known as *Sargassum*, which floats in the North Atlantic; the deadly Portuguese man o' war, whose tentacles can stretch hundreds of metres in length; and delicate pulsing blue jellies. Voracious snails and sea slugs drift over the surface, as do the ocean's only insects and one of its most misfit barnacles. Most of the species that live permanently at the surface are pigmented blue, despite the fact that blue is, generally, one of the rarest colours to see among animals. Scientists have long debated the reason for this coloration. It may be a form of camouflage to blend into the ocean's surface, or a kind of pigmented sunscreen. The second group of organisms to reside at the surface are the miscroscopic neuston, bacteria and algae capable of changing the sea from blue to gold to silver-white. These microbes may also play an important role in gas exchange between the ocean and atmosphere. Lastly, there are the rafters and the debris they ride. Rafting species attach to or shelter under floating objects. Until relatively recently, these floating objects consisted of either woody plant debris or lightweight volcanic rocks known as pumice. Thanks to human activity, floating objects now include plastics and other anthropogenic debris.

Surface species are critical to the ocean ecosystem. Many are top predators, both literally and figuratively. Although species at the surface may not look as threatening as a shark or as menacing as a barracuda, many are voracious predators of fish, fish eggs and invertebrates, often able to feed on animals many times their own size. Surface species are themselves preyed upon by diverse open-ocean predators. In the North Pacific, loggerhead turtles and Laysan albatrosses feast on floating life.

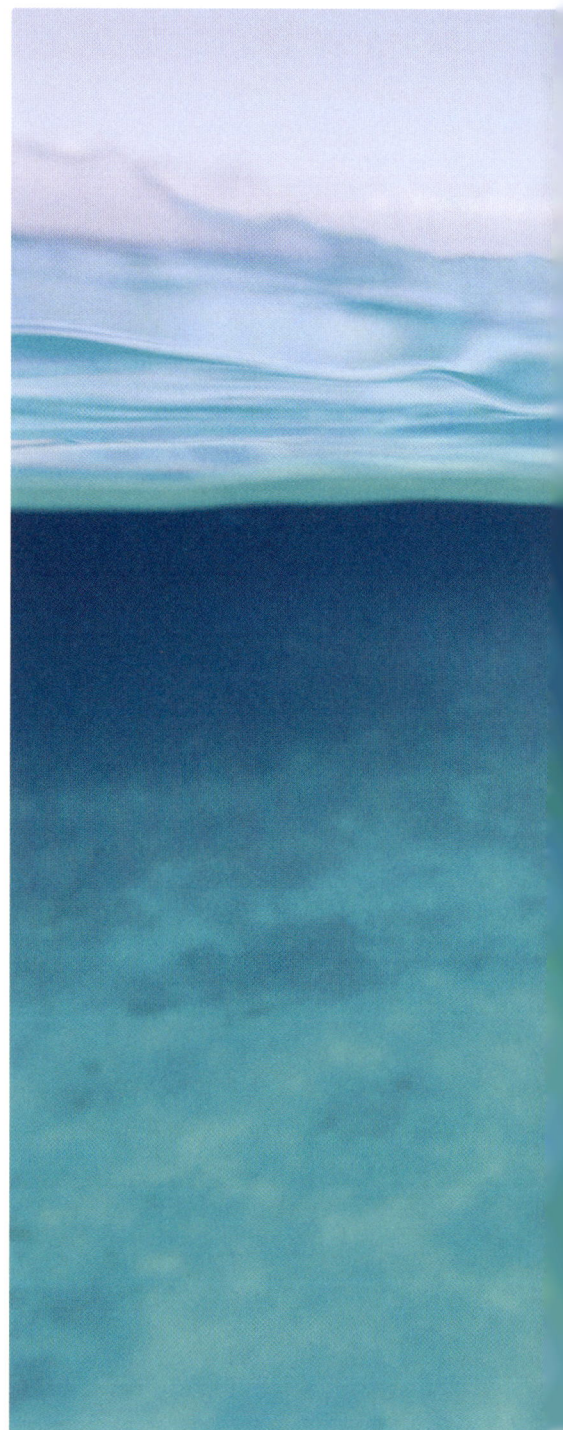

Most of the species that live permanently at the surface are pigmented blue, despite the fact that blue is, generally, one of the rarest colours to see among animals.

Portuguese man o' war
Physalia physalis
The Portuguese man o' war has a buoyant float that keeps it at the water's surface and which it also uses as a sail.

These predators may be fortunate to come across giant rafts of floating life that blanket the ocean's surface to create the illusion of living islands; at other times, they may have to hunt for animals that can be scattered by the wind and waves over hundreds or thousands of kilometres. Predators of surface life, such as turtles and albatrosses, are vulnerable to new forms of human pollution. Many are also susceptible to accidental ingestion of plastics. But plastic is not the only thing impacting the surface.

The neustonic zone is on the front line of our changing world. Climate change is rapidly altering ecosystems across our planet, and shifts that occur above the waves must pass through the surface before their impact is felt in the waters below. Likewise, a higher percentage of carbon dioxide in the atmosphere will increase ocean acidity, and this heat-trapping gas must pass through the surface to reach deeper waters. Oil spills can be paper thin and spread across the surface for hundreds of kilometres. Shipping and the associated noise pollution are all concentrated at the surface. Even light pollution from shore and ships will reach the surface first and be most intense at this air–sea boundary.

The ocean covers more than 70 per cent of Earth's surface, and diverse surface species occupy this entire region.

Changes in the surface habitat will, likewise, have a ripple effect throughout ocean ecosystems. The ocean covers more than 70 per cent of Earth's surface, and diverse surface species occupy this entire region. It is the single largest continuous habitat on Earth, and humans have already changed it in both visible and less obvious ways. For example, more wood may have entered the ocean several hundred years ago, and this wood likely floated at the surface. Now, with widespread damming and logging, the volumes of wood entering the ocean have decreased. In this way, we have had and continue to have an outsized impact on this fragile world.

The surface is at once a nursery, a hunting ground and an escape plan. This vast watery plain makes up a unique and critical habitat, full of mysteries that have preoccupied the human imagination for hundreds of years.

01 – Humpback whale
Megaptera novaeangliae
Parasitic acorn barnacles are attached to this whale.

02 – Floating blue buttons
Porpita porpita
A ring of blue tentacles surrounds a central white float.

03 – Blue sea dragon
Glaucus
Shiny blue-grey bodies help the creature to blend into the ocean's surface.

03

Sea of Gold

In 1492, as Christopher Columbus sailed across the Atlantic Ocean, he noted great mats of drifting golden matter. Hauling some of the weed on deck, the sailors noticed that it was covered in small gas-filled floats that clustered between its long slim leaves. This algae, with its small floats, resembled the tiny sarga grapes that grew in Portugal, and so it became known as *Sargassum*. The region of the ocean where *Sargassum* is found soon became known as the *Mar di Sargasso*. And through this unlikely series of less-than-notable historical events, Columbus made his only true geographic discovery. Not the Americas, which were of course quite populated, but instead the Sargasso Sea.

The designation of 'sea' might seem odd at first – this region is bound up by open ocean on all sides. But the large mats of floating algae that concentrate here set this ocean region apart from the broader North Atlantic. Golden floating *Sargassum* pool here with the help of large current systems that border the Sargasso Sea on all sides. To the west, there is the powerful North American Gulf Stream, which speeds past Florida and deflects near the Carolinas on its way to the European continent. To the east is the thick Canary Current sliding southwards past Europe and then merging into currents that carry water to the west near the tropics. The comparatively still water between all these powerful currents is known asa 'gyre', a place where floating objects, including algae, collect.

In recent years, the abundance of the Sargasso Sea has spread to other ocean regions. Beginning in 2011, scientists imaging the ocean via satellite noticed that the currents south of the Sargasso Sea were growing richer in the once-rare algae. Now known as the Great Atlantic Sargassum Belt, this region stretches from West Africa to the Gulf of Mexico, covering an area larger than the continental United States. No one knows why the region has formed, or the impact it might have on the broader ocean. Some speculate that human activity, perhaps related to fertilizer use or climate change, may be enhancing *Sargassum* growth in these regions. It is also possible that this phenomenon is part of non-human Earth processes, strange cycles or variations in Amazon river discharge or upwelling off West Africa that we have yet to fully untangle. Whatever the reason, the algae is now washing up on Mexican, Caribbean and US beaches at densities never documented before. This has raised the alarm for many.

However, our knowledge of *Sargassum* is still in its relative infancy. We know *Sargassum* was washing ashore in Mexico long before Europeans arrived. The Maya name for *Sargassum* is *U tail kaknab*, translated to 'is thrown by the lady of the sea'. From Mayan times to Columbus to the modern day, the beautiful algae is an enduring source of wonder and mystery.

01

02

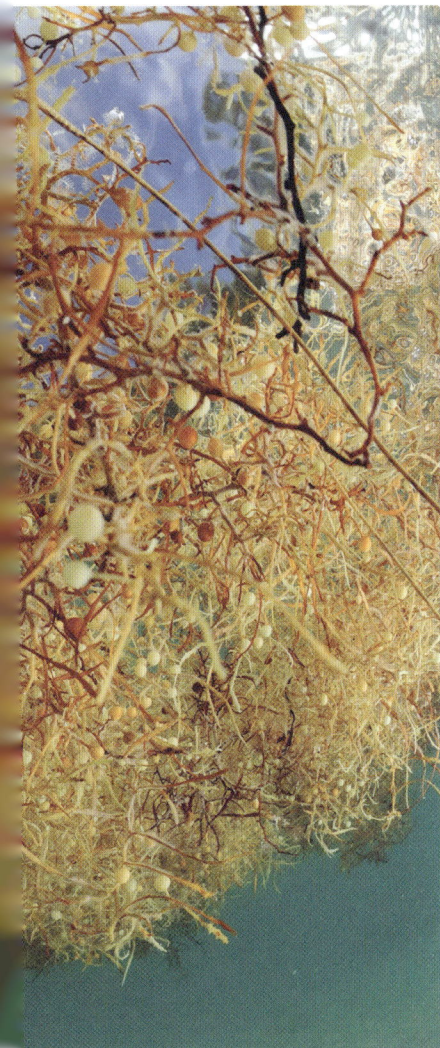

01 – Common *Sargassum* weed
Sargassum natans
Underwater, Sargasso Sea,
Bermuda

02 – A drone photo of
Sargassum covering a coastal
area
Dominica

03 – *Sargassum* washed
ashore

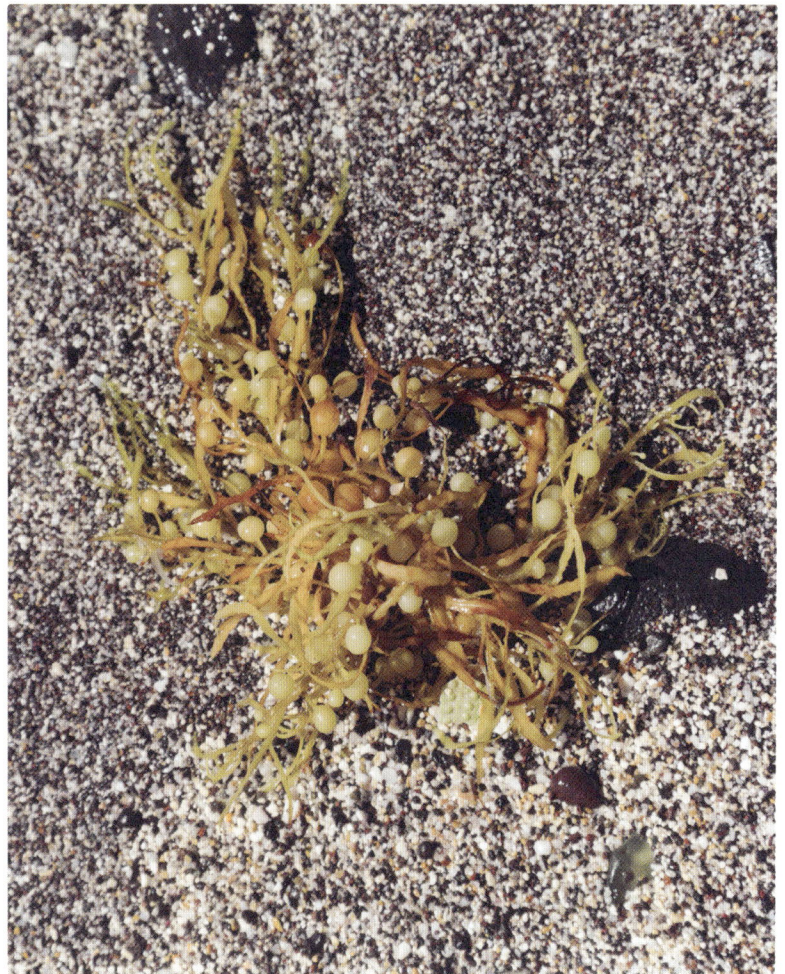

03

Man o' War

The Portuguese man o' war (*Physalia physalis*) is many things, quite literally. It is a sailing jelly that lives half in and half out of the water, a bulbous flexible gas-filled float acting as a living sail. It is a vicious and deadly predator, jigging for fish in deep water using tentacles that can grow longer than 30 m (100 ft). It is also beautiful, with a curving sinuous shape composed of violet and cerulean hues. But perhaps what is most remarkable about the Portuguese man o' war is that it is not a single living animal, but many.

Scientists have wondered for centuries about the meaning of the 'individual'. The Portuguese man o' war complicates this question considerably (see p. 162). The man o' war is nested within a group of animals known as hydrozoans, which are in turn related to corals and true jellyfish. Hydrozoans can live as solitary individuals or in colonies. In a solitary form, a hydrozoan attaches to the seafloor, no bigger than a poppy seed, with a tree-stump-shaped body capped by a whirl of tentacles around a central mouth. When food is plentiful and the water warm and suitable, a solitary hydrozoan may go through a process called 'budding', where a small mound forms on it. As this bud gets larger, it grows its own tentacles around a small, puckered mouth. The bud will then sever from the parent and settle to the seafloor. And thus a second hydrozoan individual is produced. In colonial hydrozoans, these buds remain attached to the parent polyp, creating a kind of tiny living animal tree. Thus, one animal becomes many, but each animal in the colony still retains its near full identity.

The man o' war goes one step further. It begins as a single individual, a small larva floating through the ocean, a mouth and tentacle at one end, and at the other a small bubble of gas – destined to become the man o' war's float. A small bud forms along its side, but instead of becoming a full unique connected copy of the parent polyp, the bud develops only into a part of an animal. The bud forms a whole body, but which may have only a mouth but no tentacles. Another bud may grow a tentacle but no mouth. Different individuals within the colony are like the cells of our body, each performing only one specific function. This has earned the Portuguese man o' war a special name. It is not a single organism: instead scientists call it a superorganism, a creature made up of many differentiated clones.

What we know for certain is that the man o' war is deadly. Despite its fragile jelly body, the man o' war preys on large, fast-moving and powerful fish. Its venom is one of the most potent ever described; it is used to instantly paralyse and disable stronger fish, but is also harmful to humans. With a few swipes, a man o' war can ruin a trip to the beach and even send victims to hospital.

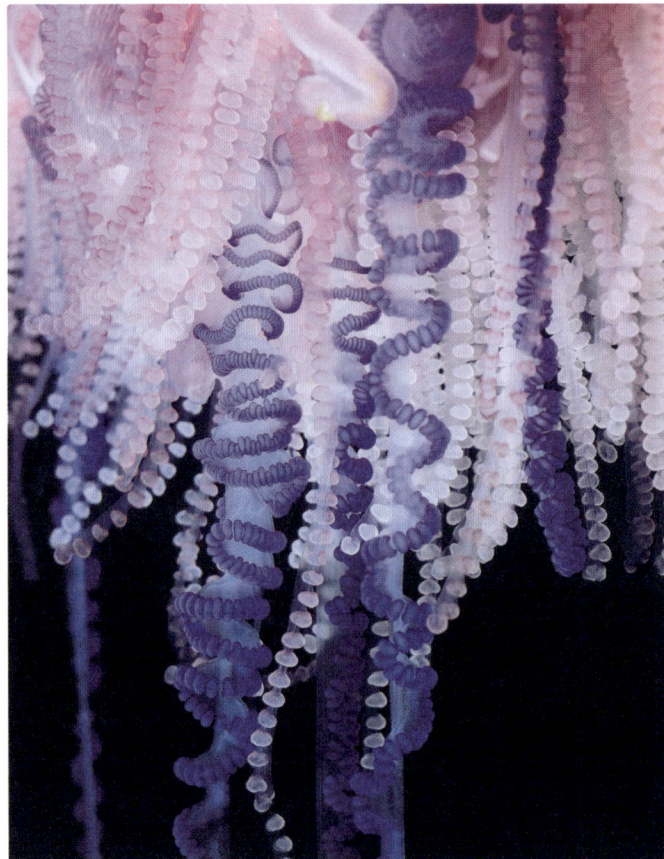

Portuguese man o' war: close-up of tentacles

Pneumatophore
Gas-filled float that plays a role in positive buoyancy, used as a sail to catch the wind.

Gastrozooid
Feeding polyp that ingests food, and secretes enzymes to break it down and digest it.

Tentacular palpon
Tentacle-bearing polyp, unique to this species. The tentacle is used to capture prey, and is specialized for nematocyst (stinging cell) production.

Snails and Sea Dragons

At the ocean's surface, the most ruthless predators have rows upon rows of needle-sharp teeth, can strike with lightning speed and hold on to prey with curling venomous tentacles. Meet the deadly snails and slugs.

The violet snail, *Janthina*, and the blue sea dragon, *Glaucus*, hang from the ocean's surface as a bat hangs from the roof of a cave. To them, they are not so much floating as they are crawling on the surface of the sky. Like the Portuguese man o' war (see pp. 68–69) they are at the mercy of the wind and currents, drifting. Waiting.

The violet snail clings to a raft of bubbles, made by dipping its purple body into the air and wrapping its muscular snail foot gently around an air pocket. It then wraps the pocket in snail slime, creating a bubble that quickly hardens into a plastic-like consistency. It affixes this bubble to its raft, but the bubble will only last several days before dissolving. And this snail cannot swim. So it floats and makes bubbles. It is patient because it has to be – violet snails have no tail to flutter and no fins to flap, so they cannot chase the food they seek. A man o' war may drift within its line of sight, but only if the currents push it close will this snail be able to feed.

The blue sea dragon has a bit more control. This slug takes large gulps of air, holding it in special cavities within its gut, located around the same area as our lungs. In bright light, these bubbles of air shine like small round gems. It stretches long handlike appendages from its body, each with many finger-like projections called cerata. The blue sea dragon can move a bit more than the violet snail: it can twist and reach and, with very slow and undulating movements, crawl along the surface of the sea just as a slug might crawl along a pavement. When swamped by waves, it can do back bends and somersaults, twisting and turning to reach the surface and land upright. Still, it cannot swim as a fish might, and so it too must wait for prey.

But when prey does come, both creatures are ready. The violet snail's head sits atop a long neck, which it tucks within its body. Two small wings flank its sides and when prey hits them, the snail lunges like a snake. Blue sea dragons do not strike, but grab. Those finger-like cerata snatch prey from the water just like a human grasping. Then both animals pull their unlucky victim in.

The mouths of these slugs and snails are gaping round toothless maws, sticky cavernous pockets that can stretch and flex to swallow massive bites. At the bottom is a tongue lined with rows and rows of cat-claw-shaped, inward-facing teeth. This tooth-covered tongue is called a radula, and snails and slugs feed by licking it across their prey, its surface acting like a living meat grinder, ripping and peeling off layer upon layer of flesh. Even the man o' war has little defence against this assault.

Blue sea dragons eat Portuguese man o' war and can steal their stinging cells

Violet snail
Janthina janthina

Blue sea dragon
Glaucus

Sargassum Dwellers

The golden forests of neustonic *Sargassum* (see pp. 66–67) provide a key environment for a wide variety of surprising species. Terns and other seabirds roost on the dense mats, gaining valuable respite during their flighted voyages across the ocean. Sea turtles nestle between fronds, using this region as a critical habitat for a developmental stage once known as the 'lost years' – the time between hatchling and adulthood when sea turtles would mysteriously disappear to a region that was, for many decades, unknown.

The Sargasso Sea's most unusual resident may be the critically endangered European eel (*Anguilla anguilla*). For centuries this eel, which can be found in rivers and ponds in Europe and the United Kingdom, was used as a form of currency. British serfs paid landowners in eel catch. And for just as long no one knew where the eels came from. Many assumed they arose spontaneously from the mud, but we now know that the eels swim down rivers and transit the vast Atlantic to reproduce under beds of *Sargassum*.

In addition to birds, turtles and eels, at least ten species can be found only on *Sargassum*, including *Sargassum*-shaped seahorses, sea slugs and frog fish, all exquisitely adapted to blend into the canopy of golden fronds.

Loggerhead sea turtle floating among *Sargassum*

Sargassum fish on *Sargassum* seeweed

Leafy sea dragon

European eel moving through seaweed

Juvenile planehead filefish hiding in *Sargassum*

A sargassum fish camouflaged in algae

Tiny crab on broad-toothed gulfweed

Buttons and Sailors

Holding a blue button (*Porpita porpita*) in your hands feels a bit like holding a living star. These creatures reside at the ocean's surface, using a circular float to anchor to the air–sea interface. Although they may be common far out at sea, they rarely wash ashore. When they do, they spill on beaches like coins tumbling from a purse, treasure from a faraway place. Cupping a pool of water with a floating blue button inside, you will see its whirl of blue tentacles twinkle and blink, while every minute or so all those shimmering tentacles will curl in and then stretch out like a flower folding back into bud and then blooming once again.

Blue buttons at shore have met their end, their natural habitat being out at sea. They are relatives of corals and attach to the ocean's surface, just as corals attach to the seafloor. Their central float is like an anchor to the air, keeping them in their preferred place at the boundary of the waves. They use their ring of blue pulsing tentacles to catch small crustacean prey. Below the float and beyond this ring of tentacles there is a forest of small white mouths waiting. Each mouth resembles a living twirling macaroni noodle. But each mouth is also small and can only manage tiny bites. And so at the very centre, above this forest of small mouths, rises the volcano-shaped central mouth, capable of swallowing the largest prey. For jellies, it is not uncommon to have many mouths, and blue buttons are no exception.

Drifting through a field of blue buttons at sea, you may also find their more industrious relatives, the by-the-wind sailors (*Velella velella*). They are like blue buttons, distinguishable only in shape rather than detail. Both species have a central float, a ring of blue pulsing tentacles and a forest of small mouths surrounding the large volcanic central maw. But the by-the-wind sailor is not quite round – instead its float is skewed to a more oval shape, with a remarkable adaptation perched atop. By-the-wind sailors are one of only a handful of ocean animals that have evolved to harness the power of the wind.

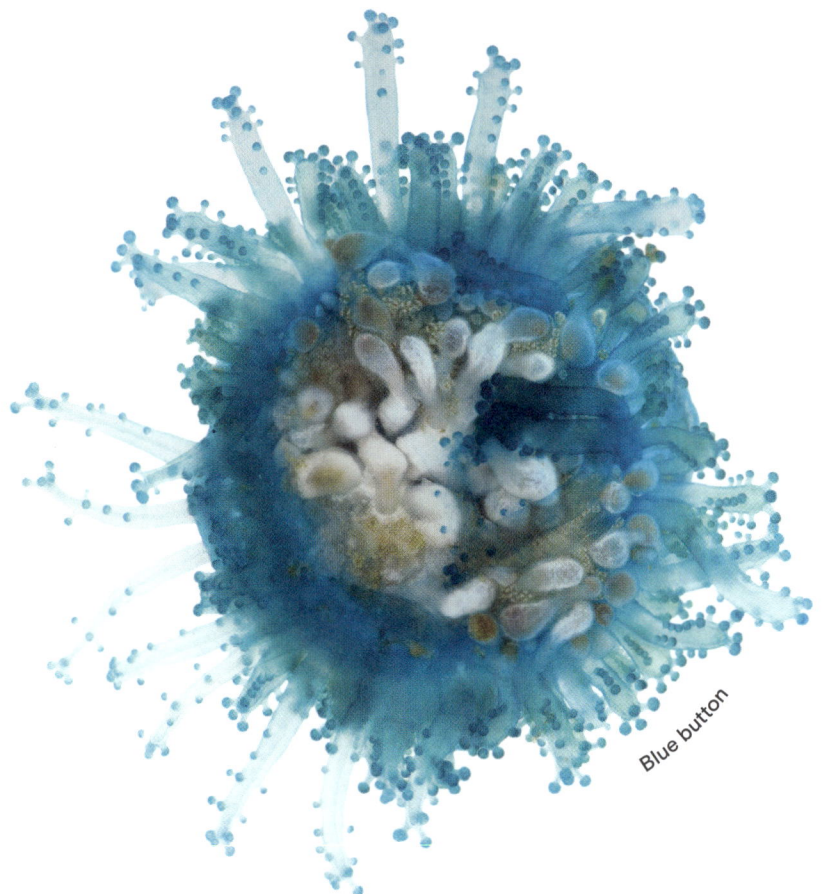

By-the-wind sailor

Blue button

By-the-wind sailor

Blue button

By-the-wind sailor

Emerging from their central float is a stiff plastic-like sail that can carry them along at speeds of roughly 6.25 m (20 ft) per minute. Unlike blue buttons, which are mostly tropical, the sail lets by-the-wind sailors travel across broad reaches of the sea, from the equator to just south of the Arctic Circle.

But both species have a problem. Whether drifting on the currents or sailing over the sea, both are subject to the whims of the world around them. This is true for most life on the ocean's surface: there is precious little control. For this reason, like many neustonic species, blue buttons and by-the-wind sailors shed their young into the deep. Both species bud near-microscopic jellyfish from the forest of their mouths, each growing as a fruit might grow on a tree. Like fruit, these jellies eventually fall from their parent animal and sink into the calmer waters of the sub-surface sea. There, they release eggs or sperm and the young blue buttons and by-the-wind sailors that form grow small bubble-filled floats and rise to the surface to start the cycle anew.

When the Sea Changes Colour

In sunlight the ocean is supposed to be blue; in darkness it is supposed to look black. Blue may shift to shades of green in shallow or nutrient-rich waters, while the black may look cool and grey under moonlight. But we do not expect these general rules to change. Microbes have other plans.

Trichodesmium lives at the ocean's surface and turns daylight water gold. These microscopic algae often live just below the surface, harvesting sunlight and mixing it with nutrients to create energy and ultimately more algae. Some nutrients, even those commonly found on land, such as iron, are rare in the open ocean, so sun-loving algae must fight for every scrap. But when conditions are just right and *Trichodesmium* have all they need, their populations explode.

Billions of algal cells can change the water from dark blue to umbers and ochres, gilding calm patches of water in a deep yellow hue, giving them the name 'sea straw'. These blooms of algae may form a continuous sheet, or stripe the surface in a garish display. It is unclear what the larger ecosystem impacts of these blooms might be, if any. They are not always easy to find or predict, so sailors come upon them at random and drift from them just as quickly.

Golden water is a wondrous curiosity, but glowing seas are shocking. This phenomenon, known as 'milky seas', is different from the more commonplace forms of light one might see at sea. Diverse ocean life forms create their own light in a process called bioluminescence. Most of the time, these creatures will flash their light only occasionally, perhaps when startled or distressed. The phenomenon of *milky sea* is different.

For reasons that are not well understood, a species of bacteria, likely a kind of *Vibrio*, will form a layer right at the surface. This layer can stretch for kilometres in all directions, reaching 17,700 km^2 (6,800 sq. miles). Then, unprovoked, they all begin to glow. The whole ocean will appear stuffed with light, a single uniform luminous radiance emanating from what should be the black waters of dark night. Milky seas can outshine the sky, creating the alarming sensation that the sky is below and the black ocean far above. The phenomenon

is extremely rare, but in 2005 scientists confirmed that a milky sea had been observed from space for the first time. It spread over dozens of kilometres and was sighted by at least one ship that transited through the region that night.

The ocean is vast and colour changes are rare. For both the golden ocean and milky sea phenomena, satellite observations may help us finally understand when and where they occur, and why gold and glowing creatures sometimes bloom.

Quote from ship that passed through milky seas, published in a scientific study:
'25 January 1995. At 1800 [GMT] (2100 local time) on a clear moonless night while 150 [nautical] mile[s] east of the Somalian coast, a whitish glow was observed on the horizon and, after 15 minutes of steaming, the ship was completely surrounded by a sea of milky-white color with a fairly uniform luminescence. The bioluminescence appeared to cover the entire sea area, from horizon to horizon…and it appeared as though the ship was sailing over a field of snow or gliding over the clouds…. The bow waves and the wake appeared blackish in color, and thick black patches of oil were passing by. Later, the Aldis lamp revealed that the "oil patches" were actually light green kelp, amazingly black against the white water.'

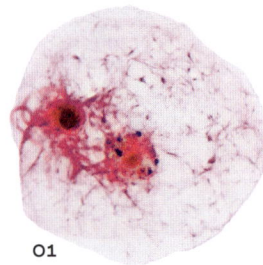

01

01 – Sea sparkles create a shimmering, though temporary, bioluminescence at the surface

02 – Marine plankton
Pyrocystis noctiluca
Similar species cause a more ephemeral form of ocean flashing, which can be seen as a sparkle in the waves.

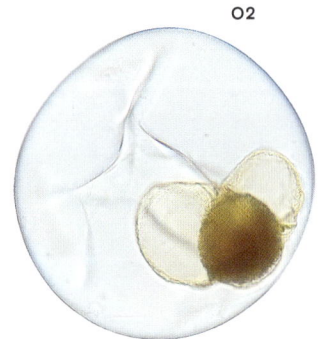

02

03 – Aerial cyanobacteria
Trichodesmium
Bloom off Queensland coast, Australia

Skating on the Sea

Of the 7.77 million species of animals on the planet, roughly 1 million are insects, and only five of those species live in the ocean. Insects are descended from marine arthropods, cousins to crabs, lobsters and other crustaceans. They emerged from the sea millions of years ago, and it is a mystery why so few have returned. Sea skaters (*Halobates*) are the open ocean's only true insect residents, spending their whole lives thousands of kilometres from shore. Like their pond-skater relatives, they glide atop the water's surface like small ice skaters, cresting over waves and surfing down their slopes. But these insects have paid a high price to make the ocean their exclusive playground.

Sea skaters begin life in eggs that must be laid on a hard surface, a constraint that many insect species face. But the seafloor may be thousands of metres below, and land thousands of kilometres away. Sea skaters must get creative. Bird feathers and driftwood make ideal surfaces, but these are few and far between. Another option is to find swimming ocean snails that venture close to the surface, and deposit their eggs on their shells. Eggs may travel hundreds of metres deep as they ride with their snail host before hatching. And when they hatch, the young face another major challenge.

Sea skater young must somehow break through the thick and sticky surface layer in order to reach fresh air. Once at the surface, waves crash and tumble, and despite their remarkable speed, sea skaters of all ages are vulnerable to being capsized. So they possess a bristly coat of long whisker-like projections, and when a sea skater is submerged, air trapped between these whiskers transforms into a built-in life jacket, encasing their body in a bubble and buoying them back to the surface.

Sea skaters are some of the fastest predators at the ocean's surface. No larger than peppercorns, they can reach speeds of 1 m/s (approx. 3 fps), giving chase to helpless drifting prey such as by-the-wind sailors (see pp. 76–77), piercing their flesh with sharp jaws and sucking out the juices. Drifting plankton and snails are also choice foodstuffs.

Life may be getting easier for these skaters. Although the abundance of driftwood in the ocean has plummeted in the last century thanks to logging and river damming, plastics are on the rise. In places where there is more plastic, there are more sea skater eggs – and more sea skaters. Perhaps these marine animals will be one of the few species to benefit from our increasingly plastic-rich waters.

Walking on water
Long pairs of rear legs spread the sea skater's weight across a wide area to retain the surface tension of the water.

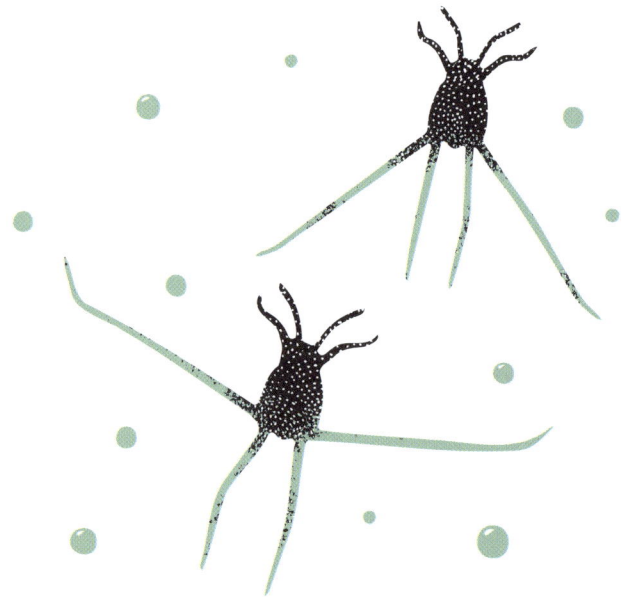

Back-up oxygen
A series of hairs and bumps traps a thin layer of air against the body of the sea skater, meaning the insects can breathe if they accidentally get trapped beneath the water and must swim back to the surface.

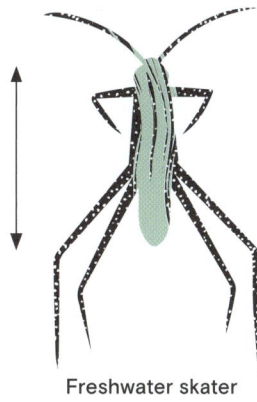

9.5 mm (⅜ in)

Freshwater skater

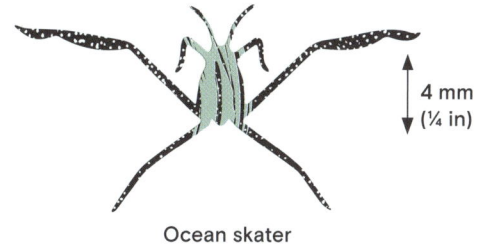

4 mm (¼ in)

Ocean skater

Agility
Strong and muscular back legs mean sea skaters can dodge, jump and somersault with ease.

Size
Ocean skaters are smaller than their freshwater counterparts, which enables faster acceleration, higher speeds and impressive reaction times.

Jumping
Sea skaters use their body weight to push down on the water, which creates a 'bend' in the water surface. Extreme acceleration allows rapid escape from aerial and underwater predators.

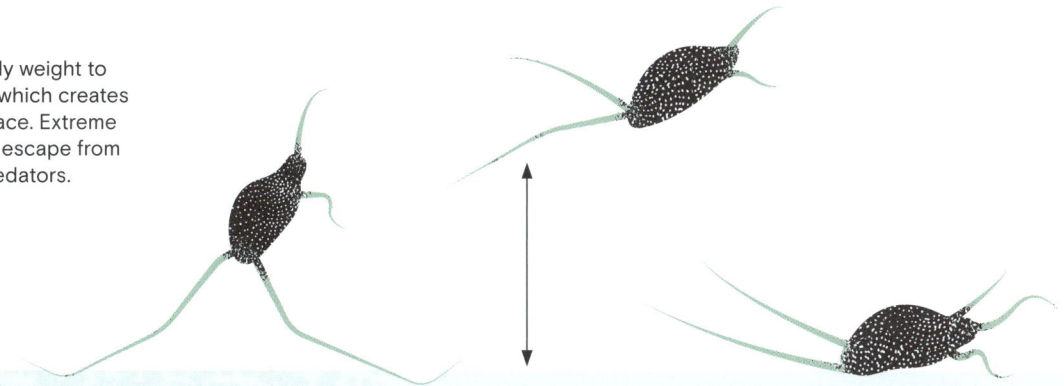

Bending of water surface

Bounces off the water surface

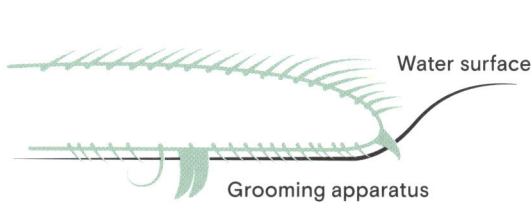

Water surface

Grooming apparatus

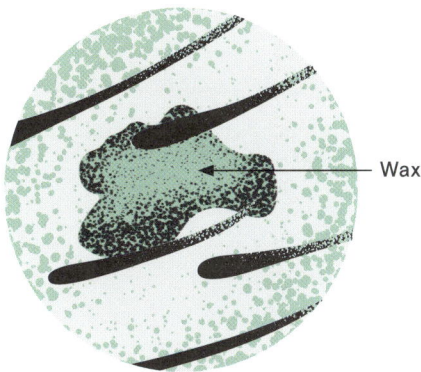

Wax

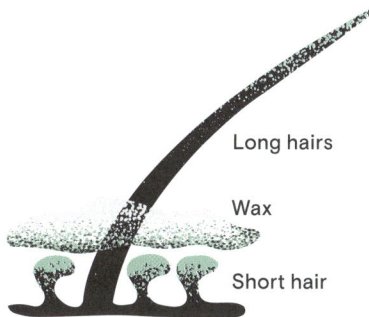

Long hairs

Wax

Short hair

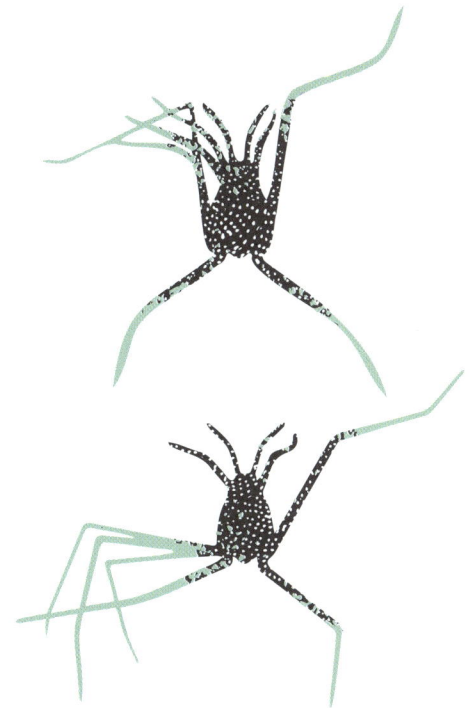

Superhydrophobicity
Specialized body hair combined with a wax coating gives sea skaters multilevel hydrophobicity (meaning they are extremely water-repellent). During grooming, the insects transfer the secreted waxy coating from the abdomen to the rest of the body and the legs.

Flying above the Waves

A flying fish can cover more than 400 m (1,300 ft) in thirty seconds before slipping back under the surface. Fish are not the only ocean species to take to the skies: squid, copepods (tiny crustaceans) and sea skaters are all known to use the air as a place of retreat.

Flying fish compose a large family of species, all of which share the same wing-shaped pectoral fins. These fins are tucked tight to the fish's slender body when they are swimming, but provide a lifeline in times of danger. If pursued by a predator from below, these fish beat their tails towards the surface and launch themselves from the water, their large pectoral fins spreading wide to act as gliding wings. Sailing is such a critical adaption that the young are born with colourful butterfly-like wings and can take to the air at only a few weeks old.

Flying fish have a long history in sailing lore, but flying squid are an oddity. Squid (including, but not limited to, the Japanese neon flying squid) provide a delicious meal for both human and non-human animals. With no hard shell or poisonous venom to protect themselves, squid are tasty swimming protein bars. So they must be smart and they must be fast. When the flying squid senses a threat, it sucks water into its cavernous body and rockets it through a siphon, generating the jet force needed for liftoff. Once it has broken the surface, the squid splays its tentacles and fins into a winglike surface, creating lift that allows it to soar for up to 55 m (180 ft). Squid are much closer than many to mastering the art of flight.

Large animals might take the crown for the most impressive ocean gliders, but many organisms retreat above the waves. Rice-sized blue crustaceans called copepods cling to the air–sea boundary as a bug might cling to a leaf. Their blue colour may help camouflage them against the sky, but it is not always enough. When a predator gets too close, they launch themselves into the air, tumbling around and around like trapeze artists before eventually plopping back into the water. Sea skaters are also capable of impressive jumps. More like ice skaters than circus artists, they pull their legs tight in an acrobatic display.

Such aerial feats make catching these species tricky for both marine predators and scientists. Many a biologist has captured a beautiful blue copepod or young flying fish only to have it fling itself from the bucket and back into the sea.

01

01 – Copepod

02 – Flying fish have winglike fins to sail through the air

03 – Water strider insect

04 – Neon flying squid
Ommastrephes bartramii

02

03

04

THE SURFACE

Gathering at the Surface

Compared to water even a few metres below, there is often an abundance of life crowding the air–sea interface.

The surface is a nursery ground and meeting place for species from the deep sea to coral reefs to coastal zones and even, in the case of eels, fresh water. While precious little research has been done on why so many species meet here, we can guess. The surface is both a shelter and a social hub.

Coordinating a meeting in the vast three-dimensional expanse of the water column requires each party to know where the other will be on three different planes. You must know how deep they are, but also their latitude and longitude. The surface, in contrast, reduces this number to two. Thus many species may use the surface as a sort of landmark, a way to find each other. Hundreds of fish species have floating eggs; these eggs may be more likely to bump into fertilizing sperm in this thin layer. More than 100 fish species are documented to spend their juvenile days here, including seahorses, swordfish and mahi-mahi.

In addition to its convenience, the surface also offers a unique suite of things to eat. On top of the true neuston, insects or other debris from land may be swept out to sea and eventually rain onto the surface. Beetles and bees are not uncommon to collect from surface waters hundreds of kilometres from shore. Floating seagrass and tree seeds may also be found, carrying with them potentially delectable coastal creatures. Currents and wind at the surface also concentrate life from sub-surface water into dense regions called slicks. These slicks may comprise as little as 8 per cent of the surface by area, looking like smooth stripes of water when viewed from above, but they may contain nearly 40 per cent of larval invertebrates and more than 90 per cent of young fish, probably thanks to the abundance of small food that can be found within them. These patches of bounty provide the means for small creatures to collect food without exerting as much energy as they would if they foraged in deeper waters.

The surface may also provide safety. In the open ocean there is no place to hide: a hunting predator can see you from any angle. Just as a fearful human caught in an open space might cling to the safety of a wall – something, anything, to provide some cover – diverse ocean life may seek out the surface as a kind of shelter. And when there is an imminent threat from below, it provides an escape.

03

04

05

01 – Flying fish
Exocoetidae

02 – School of bright-blue needlefish
Belonidae

03 – Jellyfish
Scyphozoa

04 – Needlefish
Belonidae

05 – Red Sea houndfish
Tylosurus choram

Rafting Species

Trees have an afterlife at sea. Logs carried from forests on fast-flowing rivers or trees felled near shore may ultimately be swept into the ocean, where whole ecosystems are waiting to hide among their waterlogged branches.

Before industrial-scale damming and logging, there were many more trees flowing from the shore, and diverse species adapted over millions of years to rely on this driftwood (see also pp. 30–31). Now-extinct sea lily species (animals related to sea urchins and starfish) have been found fossilized within their floating wooden homes, frozen in rock for more than 180 million years. The sea lilies reached over 20 m (65 ft) long (compared to modern sea lilies, which may only muster a metre or so). Forests and oceans have been linked for nearly a quarter of a billion years. And despite the decline in woody debris due to human activity, species at the surface still depend on trees.

At least six species of clam can be found only in wood floating at sea. These clams, commonly referred to by the misnomer 'ship worms', became a central focus of human attention during the European Age of Exploration (fifteenth–seventeenth centuries). They bored into the hulls of ships and, using special symbiotic bacteria in their gut, consumed and digested the wood as they grew. Their activity quickly rotted the ship's planks, making them prone to snapping and risking the lives of all aboard. Yet the situation is just as precarious for the clam. Once settled into a single piece of wood, they cannot move to another, and so they are destined to eat themselves out of house and home and eventually sink with their decaying dwelling into the deep. But they are not the only creatures who depend on wood.

Before the eventual sinking and dissolution of all wood at sea, it is home to dozens of invertebrate species and large schools of fish. The reason for this aggregation of open-ocean life around floating debris is unclear, but hundreds of skipjack tuna have been documented around a single floating log. People of the island of Tobi, Palau, in the western Pacific, are among the first documented to fish around floating trees.

The practice was a well-kept local secret that yielded remarkable results. But in the 1990s the idea reached the global stage.

Now, diverse fisheries intentionally throw floating debris into the sea, attempting to recreate the reeflike community that, for millions of years, grew on trees. Nearly 40 per cent of all tuna are caught with these so-called 'fish-aggregating devices' (FADs). But they are no longer made of wood: these days most FADs are made of plastic. Species such as ship worms cannot eat them, and so as wood at sea declines, they have nowhere else to go. The species that can live on plastic now find themselves as strange and unlikely refugees clinging to plastic FADs, destined to be scooped up with the fish or left to drift beneath them. Some FADs may be as small as a few barrels tied together; others are massive, weighing several thousand kilos. Estimates suggest that around 121,000 fish-aggregating devices are launched each year in world's oceans. An estimated 33 per cent of them may never be recovered.

02

01

01 – Crinoid (sea lily) fossil
Platycrinites gigas

02 – Crinoid (sea lily) fossil
Uintacrinus socialis

03 – Crinoid (sea lily) fossil
Uintacrinus socialis

04 – Fish gathering at a
'fish-aggregating device'

05 – Crinoid (sea lily) fossil
Scyphocrinus elegans

04

05

03

Surface Hotspots

The ocean collectively weighs roughly 1.37 billion km³ (330 million cu miles). All this water is being churned and turned by our spinning planet, the pull of our moon and the prevailing winds. This mixing creates regions of our seas that serve as sanctuaries for floating life. The Sargasso Sea (see pp. 66–67) is one such place, but there are others.

Scientists have described only one neuston sea – the Sargasso – with any degree of certainty, but we have clues about where the others might be, as their locations are littered with rubbish. These so-called 'garbage patches' are areas of the ocean where plastic debris collects. One such region, known as the North Pacific High or, colloquially, the Great Pacific Garbage Patch, has been confirmed and studied for several decades.

The Great Pacific Garbage Patch sounds dramatic, but when you take a ship to the middle of its densest region and stand atop the deck, this is what you see: ocean. In fact, the density of plastic in this region is less than on many popular tourist beaches. What makes it remarkable is its distance from shore. Plastics from around the North Pacific (mostly fishing debris) are swirled into this region by currents, just as the Sargasso Sea concentrates floating algae. And, for a long time, plastic is all that was known from this region. Until a man named Ben Lecomte decided to swim across it in 2019.

Lecomte planned his journey with a team of scientists who helped him map a path through the densest part of the so-called garbage patch.

How long until it is gone?

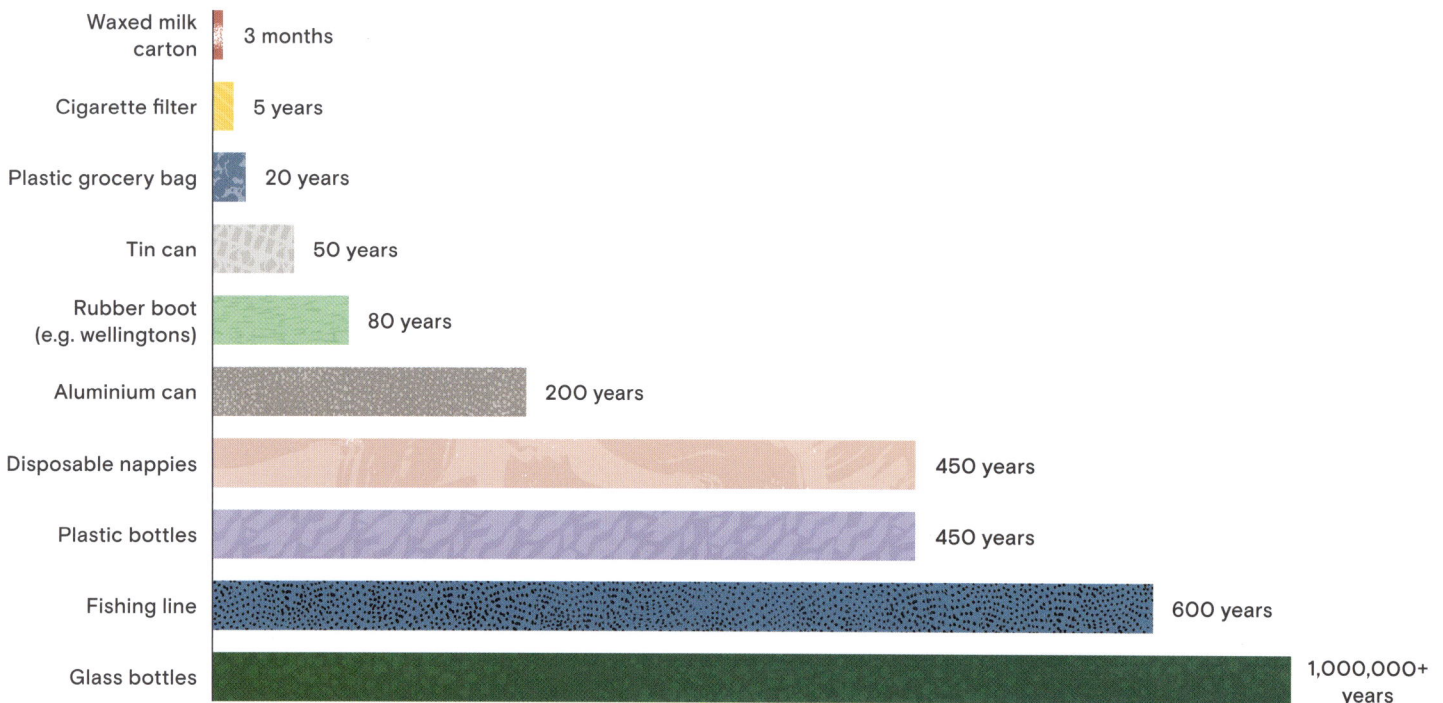

Waxed milk carton	3 months
Cigarette filter	5 years
Plastic grocery bag	20 years
Tin can	50 years
Rubber boot (e.g. wellingtons)	80 years
Aluminium can	200 years
Disposable nappies	450 years
Plastic bottles	450 years
Fishing line	600 years
Glass bottles	1,000,000+ years

While he swam, a support ship collected samples of the neuston. Until his swim, scientists had studied the types, concentrations and densities of plastic in the region, and most of the studies threw out all the neustonic life that was collected with the plastic. But as Lecomte swam, he noticed an abundance of living things in the region, later recalling that there was more life in this garbage patch than anywhere else on his swim across the Pacific. So perhaps it is not surprising that when the crew pulled up their nets, they found them full of violet snails, blue buttons and by-the-wind sailors (see pp. 70–71 and pp. 76–77). This region is so much more than a garbage patch. It is a neuston sea.

It is likely not the only hotspot. From the Gulf of Mexico to the South Atlantic, the Arabian Sea to New Zealand, scientists are now identifying other regions that may host exceptional abundances of floating life. These hotspots may help solve some of the mysteries surrounding the ocean's surface: how can predators survive if they cannot hunt? How can species reproduce if they must bump into other similar species with which to mate? These constraints have long puzzled scientists. Hotspots such as the Sargasso Sea or the North Pacific Neuston Sea may provide the answer.

Other mysteries are more persistent. What impact might plastic have on life at the surface; or the continued loss of natural woody debris? What about shipping, fishing and climate change? As with so much in the sea, the mysteries extend far deeper than they seem. Even at the ocean's surface.

Plastics collected in the Great Pacific Garbage Patch

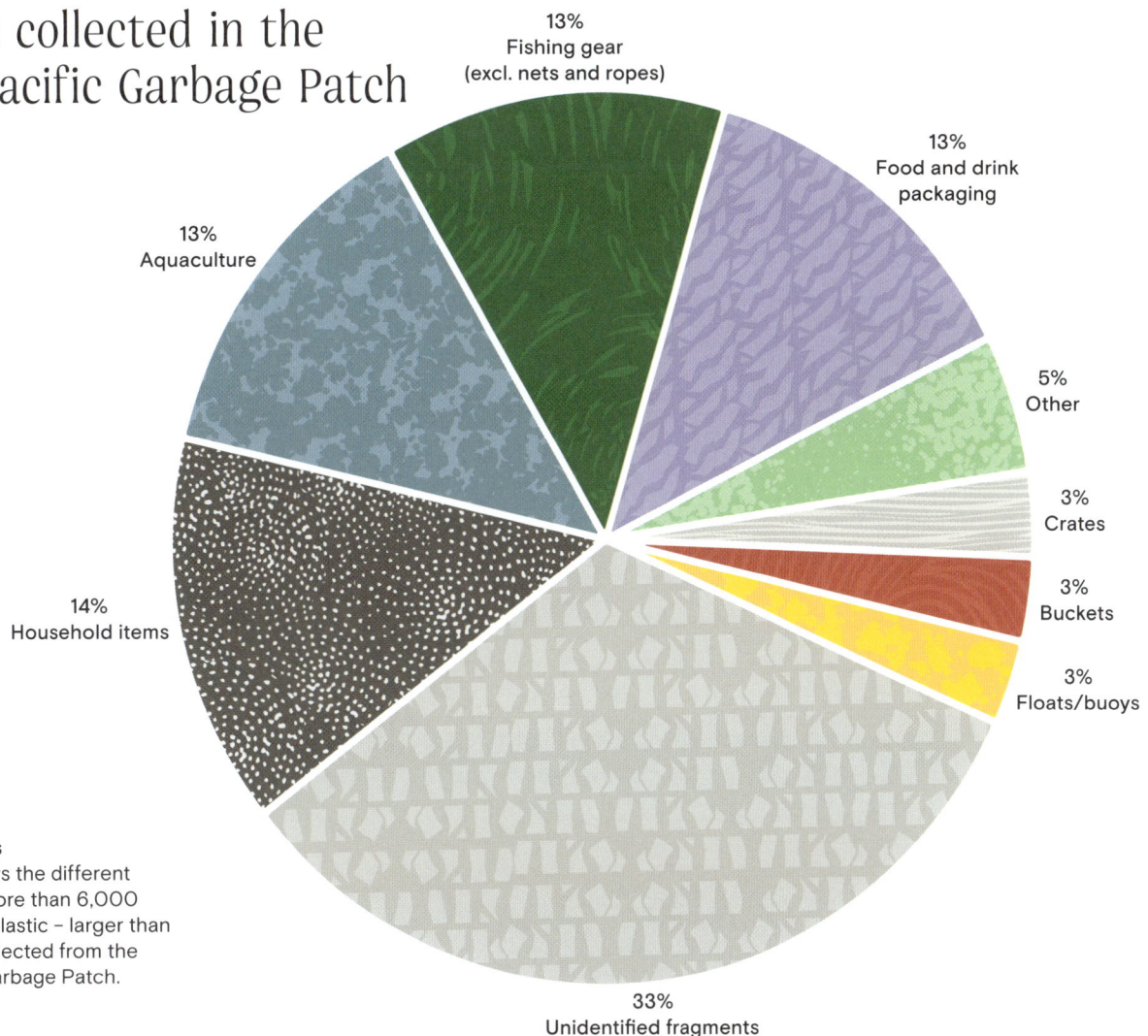

13%
Fishing gear
(excl. nets and ropes)

13%
Food and drink
packaging

13%
Aquaculture

5%
Other

3%
Crates

3%
Buckets

14%
Household items

3%
Floats/buoys

Number of items
This graph shows the different categories of more than 6,000 pieces of hard plastic – larger than 5 cm (2 in) – collected from the Great Pacific Garbage Patch.

33%
Unidentified fragments

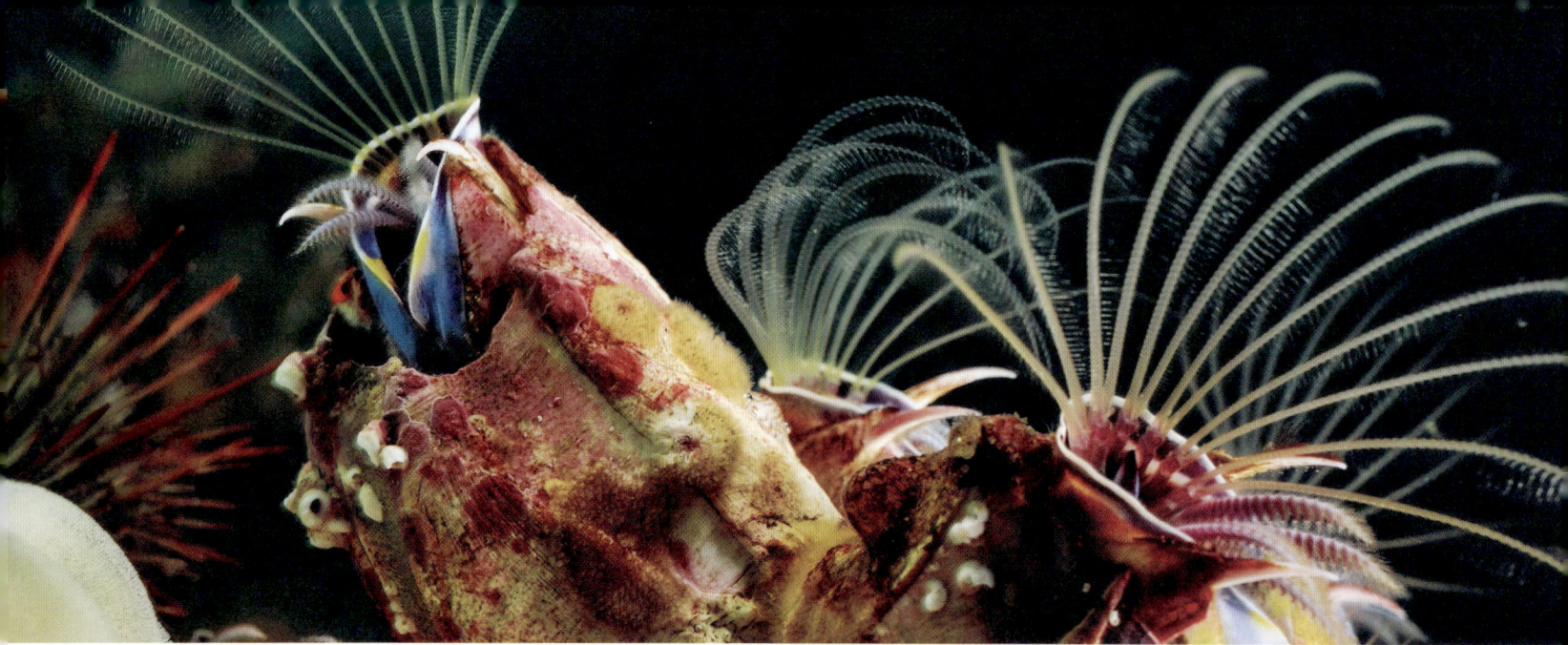

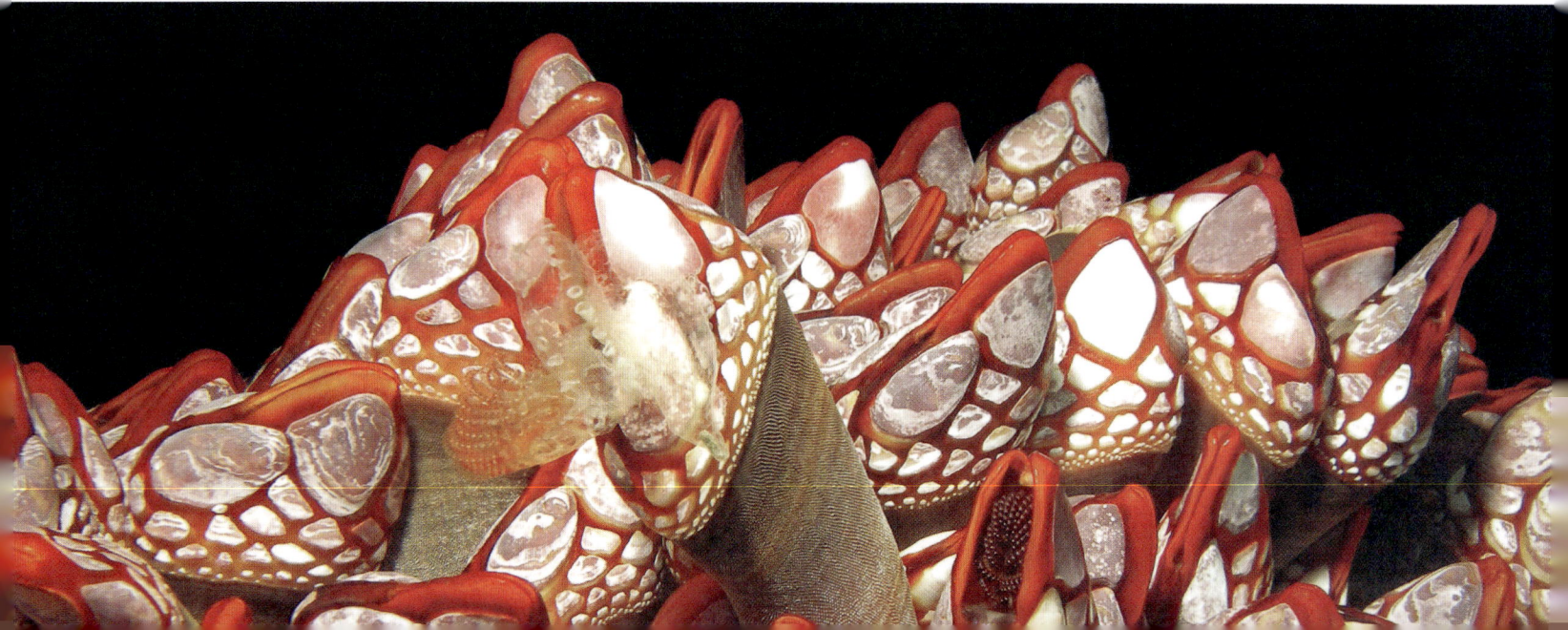

Barnacles on the Move

Though many people think of barnacles as clams, they are actually among the world's most unusual crustaceans, tucking their body inside a hard shell and using their legs to feed, giving up life in the wilder world to anchor themselves to a single permanent spot. With the remarkable exception of the buoy barnacles.

Buoy barnacles (*Dosima fascicularis*) are the black sheep of the barnacles, a wanderlust species from a family that prefers to stay rooted in place. Still, the buoy barnacle cannot escape the trappings of a barnacle: its body is tucked away inside a hinged shell. Its legs emerge, waving around to feed. It has precious little means to move around. Instead, it creates a home that can move with it.

Barnacles attach to the seafloor using a cement they ooze from their bodies. This substance acts like tile grout, securing them firmly in place. The buoy barnacle makes cement too, but instead of being tough and hard, it is more like a marshmallow. While much of nature bids for security and protection, the buoy barnacle is an example of how a species may succeed by going soft and being flexible.

Nevertheless, the buoy barnacle is not without defences. Its legs are lined with dagger-like spikes with which to catch fast-swimming prey, propelling it up the food chain compared to its filter-feeding cousins. This adaptation may be essential. As the buoy barnacle drifts into unfamiliar waters, it is able to eat exotic and novel foods.

Reproduction can prove more of a challenge. Barnacles have evolved the longest penis on Earth relative to their size; the standard shoreline barnacle penis can be eight times the length of their body. Most barnacles, anchored to rocks surrounded by other individuals, use their penis to gently probe the shells of their neighbours and, if they're keen, will mate. But what to do if you are adrift at sea? While buoy barnacles sometimes find themselves drifting around with other barnacles, the strength and length of a phallus required to procreate across the tumbling waves is a tall ask even for evolution.

While we are not aware of science documenting a buoy barnacle in the act, one answer may be the flotillas these barnacles sometimes form with other barnacles. Upwards of ten buoy barnacles may be found together on a single large float. And perhaps it is these barnacles, constructing their own loveboat, that are able to reproduce and spawn new buoy barnacles. As with so much in neuston, their secrets are their own.

04

01 – Giant barnacles
Austromegabalanus psittacus

02 – Rafting barnacles

03 – Gooseneck barnacles
Pollicipes polymerus

04 – Buoy barnacle
Dosima fascicularis
Buoy barnacles are capable of building their own floats, which they use to ride the waves.

The
Sunlight
Zone

Introduction

Our first encounter with the ocean is in the sunlight zone, the epipelagic zone. As the name suggests, this top 200 m (650 ft) is a place of light, colour and life. Although it represents just 2–3 per cent of the entire ocean, the sunlight zone is home to a staggering number of organisms.

Benefiting from its warmth and sunlight, 90 per cent of all ocean species live or spend time at this depth, including whales, sharks, jellyfish, crabs, anemones and thousands of fish. Photosynthesis, the process by which plants and algae convert sunlight into energy, is also possible in these light-strewn waters. Tiny phytoplankon – microscopic marine algae, invisible to the naked eye – use photosynthesis to produce much of the food for the whole ocean ecosystem, while simultaneously creating at least half of the oxygen in the atmosphere.

The sunlight zone offers the first glimpse of life underwater. Salinity, temperature and pressure are tolerable in these surface layers, inviting exploration. Still, the salinity, on average 35 parts per thousand (ppt) or 35 grams per litre, is significantly higher than the average for freshwater (0.5 ppt or less) and reminds us that we are entering a new and unfamiliar world. The higher salinity results in an increase in density that, in turn, makes floating at the surface easier. This zone is also relatively warm because of the heating of the sun and the constant mixing of the wind and currents. There are wide variations in temperature across this zone, dictated by both season and latitude: sea surface temperatures can range from 36°C (97°F) in the Persian Gulf to -2°C (-35.6°F) near the North Pole. Pressure increases linearly with depth – one atmosphere for every 10 m (33 ft) of descent. While this zone represents just 2–3 per cent of our ocean, it is the most accessible, hospitable and familiar to us as humans.

The sunlight zone is home to two groups of marine organisms: plankton and nekton. Plankton are named after the Greek word for 'wandering' or 'drifting'; they do not have the ability to actively swim any distance, but move passively through the ocean, free-riding on currents. Planktonic plants are called 'phytoplankton' and include diatoms, dinoflagellates and cyanobacteria, while planktonic animals are called 'zooplankton' and include jellyfish and small crustaceans. The second group comprises the pelagic (living in the open sea), free-swimming 'nekton'. This group includes marine mammals, larger crustaceans, squid, and fish.

Plankton are appropriately named after the Greek word for 'wandering' because they do not have the ability to actively swim any distance, but move passively through the ocean, free-riding on currents.

Under the surface, light streaks through, and everything takes on a blue hue. The red, yellow and green wavelengths of sunlight are absorbed by water molecules in the ocean, leaving behind the shorter-wavelength blues and violets. Sunlight rarely penetrates beyond this depth, making this the only zone where photosynthesis is possible. Photosynthesis is critical to the existence of the vast majority of life on Earth, and it is also critical to the food webs in the ocean. During photosynthesis, plants such as phytoplankton use water and carbon dioxide in the presence of sunlight to create oxygen and energy (in the form of sugar). Air breathers on land and in water use the oxygen released. There are a billion billion billion phytoplankton in the world's oceans – more than there are stars

01

The key ingredient in this zone, sunlight, provides energy and light but also makes it easier for predators to hunt and, if too intense, can make its waters uninhabitable.

04

01 – Lion's mane jellyfish
Cyanea capillata
Drifting underwater off Bonaventure Island, Quebec, Canada

02 – Fish scales: macro detail

03 – Orange sea slug
Godiva banyulensis
Off the coast of Croatia

04 – Parrotfish
Scaridae
Swimming among coral

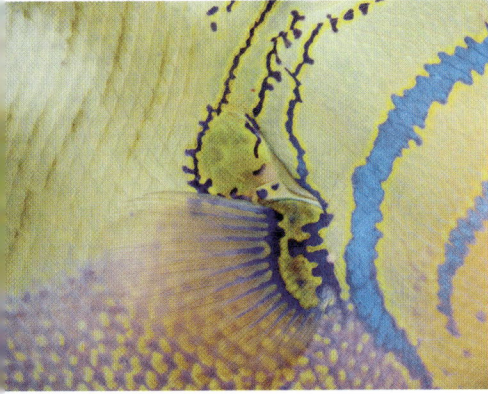

in the sky. They truly are small but mighty ocean heroes. Sixty-five per cent of all plankton are found in the top 500 m (1,600 feet) of the ocean. Still, they tend to be concentrated in areas close to land, with higher nutrient content relative to the open ocean – the latter is considered nutrient-poor, with limited phytoplankton growth. The sunlight penetrating these waters results in higher productivity and density of organisms relative to the other, deeper oceanic zones.

The ocean is not a place of silence – even less so in shallow waters, where natural and humanmade sounds are constant. Crashing waves, wind and water currents mix with the noise of organisms getting on with their daily lives. Coral reefs make a sound like popping corn thanks to the invertebrates and fish that seek refuge in their complex structures. Research has shown that noisier reefs are healthier, with the different frequencies of sound indicating the number of fish living there and the diversity of available coral.

Organisms must adapt to their environments, particularly as survival depends on their ability to find food, avoid predators and reproduce. Most predators in the sunlight zone use vision to hunt their prey. In shallower waters, organisms such as cuttlefish may change the colour and texture of their skin to blend in with a coral reef, using camouflage to evade detection. Other species may simply hide in crevices and spaces that are available. Interestingly, in the degraded and less complex reefs of the Caribbean, researchers have found that small predatory fish (such as trumpetfish) will closely shadow bigger or non-threatening species (such as parrotfish) to approach their prey without being seen. The larger fish provides motion camouflage for the smaller fish, unwittingly enabling it to capture its next meal.

As we go deeper, leaving behind the reefs and rubble of the shallows, blending into the featureless three-dimensional space of the midwater zone becomes increasingly tricky. Organisms living in this space have adapted accordingly. They become either small – making them hard to spot – like phytoplankton, or transparent, with a density close to that of water, like jellyfish. Alternatively, they camouflage themselves through 'countershading'. Countershaded fish often have a colour gradient running from dark on the dorsal (top) surface to light on the ventral surface (underside). This clever technique makes it harder to see the fish from both above and below. Becoming silvery is another similar adaptation, reflecting sunlight in the surface waters while helping the fish blend in with their surroundings elsewhere. Yet another form of adaptation, used by such fish as tuna, is to be fusiform or torpedo-shaped. This affords them superior swimming ability, enabling them to avoid predators while efficiently capturing their prey. This is particularly important for pelagic species that migrate vast distances in the open ocean for food.

Despite being the most accessible and vibrant part of our ocean, the sunlight zone has complexities. The key ingredient in this zone, sunlight, provides energy and light but also makes it easier for predators to hunt and, if too intense, can make its waters uninhabitable. Survival, hunting and reproduction methods have evolved over millennia to create the thriving metropolis that exists underwater today.

Plankton

Wunderpus octopus larvae

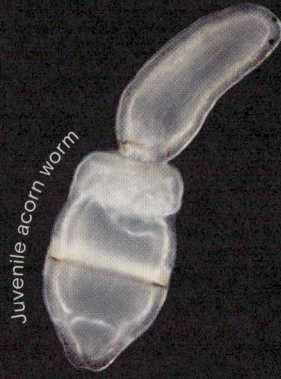

Juvenile acorn worm

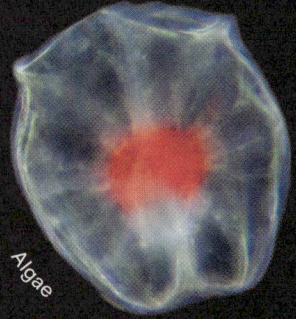

Algae

Cassiopea jellyfish

Krill

Atlantic longarm octopus larva

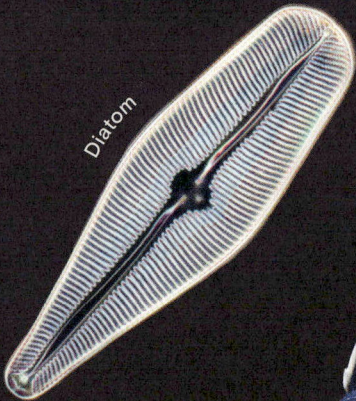

Diatom

Unidentified larval tonguefish

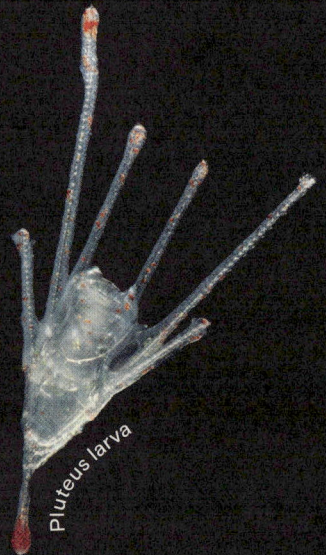

Pluteus larva

Sea sapphire

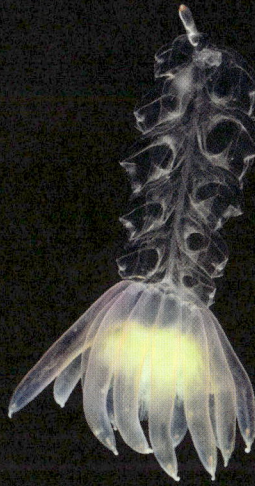

Siphonophore

Cyanobacteria

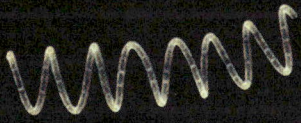

Jellyfish larva

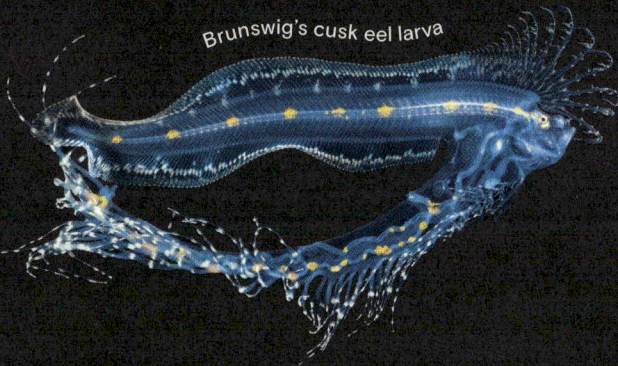

Diatom

Brunswig's cusk eel larva

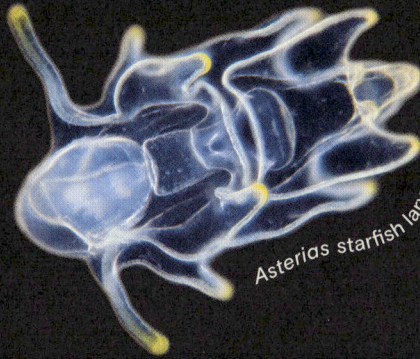

Asterias starfish larva

Shrimp

Atlanta peronii planktonic gastropod

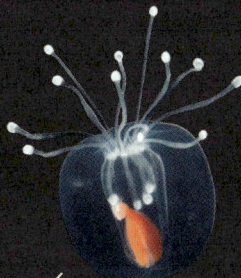

Hydromedusa

Seagrass Services

Seagrasses are the only flowering plants that can survive immersed entirely in water, where they flower and are pollinated. Although they are called seagrasses, they are more closely related to ginger and lilies than to true grasses.

Seagrasses are often found in shallow coastal waters with plenty of light, which they need for photosynthesis. They grow in sediment on the seafloor with erect, elongate leaves and a buried rootlike structure (rhizome) to keep them anchored. Seagrass meadows can comprise a single species or up to twelve species of seagrass. These immensely productive habitats provide a range of services, including vital carbon sequestration – they actually store excess carbon at a higher rate than forests. They also stabilize coastal sediment and buffer or filter chemical and nutrient inputs into the marine environment.

Seagrasses are a top-rated nursery for commercially important prawn and fish species but, most importantly, they are the single food source of the dugong – one of the original 'sirens' of the sea (see p. 253). Dugongs are called sea cows because they use their strong, cleft upper lips to graze on seagrasses they uproot from the seafloor.

Food

Seagrasses support a diverse network of oceanic wildlife, from herbivores that survive on grass and algae to carnivores and decomposers that consume dead organic matter.

Carbon Storage

Seagrasses remove carbon from seawater and use it to photosynthesize and grow. They also store carbon through sedimentation.

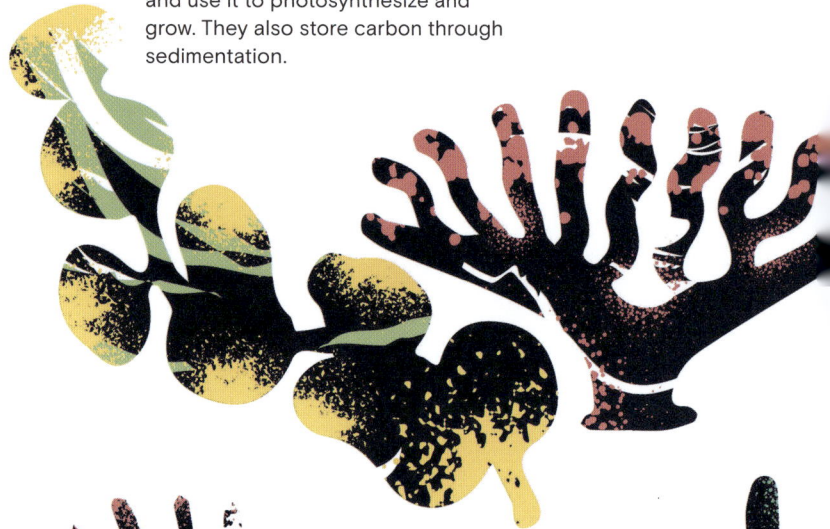

Water Quality

Seagrass roots trap and stabilize sediment, helping to improve water quality and reduce coastal erosion.

Climate Resilience

Seagrasses can survive in warmer waters, providing a critical habitat for a range of species as sea temperatures rise owing to climate change. Fisheries are likely to become increasingly reliant on seagrasses.

Habitat Corridors

Seagrass beds facilitate ecological connections between different habitats, allowing species to move between them.

Biodiversity

Thousands of marine species rely on seagrasses, including fish, marine mammals, birds and invertebrates. Several of these are endangered, such as the dugong and sea turtle.

Oxygen

Seagrasses produce oxygen through photosynthesis, which marine species need to breathe.

Livelihoods

Seagrasses are hugely important for humans all over the globe, supporting fisheries and tourism.

Nutrient Cycling

Seagrasses contribute to nutrient cycling by filtering chemicals and nutrients into the marine environment.

Shelter

Many species rely on seagrasses as nurseries to shelter their young.

Coral Coloration

Swimming through a coral reef can sometimes resemble an explosion in a paint factory, as striking fish navigate brightly coloured coral and sponges. But many corals are actually dull green or brown, so where do the bright hues come from?

To understand colour, we need to look at the drab corals first. Corals are living colonial organisms: each is composed of hundreds to hundreds of thousands of individual but genetically identical animals, called polyps. The skeleton itself is ghostly white – any colour comes from the animals that live within it. Coral polyps get some of their food through filter-feeding, but the vast majority is produced by zooxanthellae, single-celled algae that live as symbionts within their tissues. The algae photosynthesize using their chlorophyll, which gives them their green-brown colour. The vast majority of the sugars produced during photosynthesis are used by the coral polyps, with a smaller portion being used by the algae themselves. In return for the feast, corals shelter algae. This relationship benefits both species and is called mutualism.

While all corals have a green-brown hue on account of their algae tenants, they also produce protein pigments that predominantly reflect light in the purple, blue, green or red parts of the spectrum, while other proteins are fluorescent and absorb one colour of light, such as blue, and emit another colour – green or red. Coral pigments are also the algae's natural sunscreen, providing protection from harmful UV rays and increasing in quantity when exposed to very bright light to increase protection.

Acropora coral

Flowerpot coral

Wall hammer coral

Octocoral

Brain coral

Soft coral

Star coral

The Sardine Run

The KwaZulu-Natal sardine run of southern Africa is one of the greatest wildlife spectacles on the planet. Between May and July, billions of sardines – or more specifically the pilchard *Sardinops sagax* – just 25 cm (10 in) in length, spawn in the cool temperate waters off the Agulhas Bank on the southernmost tip of the continent, travelling northward along the east coast of South Africa into the subtropical waters of the Indian Ocean. This movement is triggered by the sardines from the Atlantic Ocean straying too far into southern waters and experiencing a brief pulse of cold, nutrient-rich water upwelling along the coast. These cold-adapted fish then follow the chilly currents northeast until they reach the warm waters of the Indian Ocean, where the upwelling ends and they get stranded in a challenging subtropical habitat.

This journey from familiar to unfamiliar seas, where conditions become increasingly inhospitable for them, is not an easy one. While the shoals are massive – as long as 15 km (9 miles), as wide as 3.5 km (2 miles) and as deep as 40 m (130 ft) – they also become a target. The billions of sardines do not travel alone, because where there is food, there are predators. The first to arrive are the common dolphins, tightly rounding up the sardines into 'bait balls'. Soon after, the sharks make their entrance – bronze whalers, duskies, grey nurse sharks, blacktips, spinners and Zambezis. The large game fish such as mackerel and tuna are also keen for the buffet, accompanied by the Cape fur seals and expert avian dive bombers such as Cape gannets (see p. 114–15) and penguins. Last but not least, with mouths wide open and throat pleats stretched, the humpbacks and Bryde's whales arrive. Despite the huge numbers of fish involved, the sardine run involves only a fraction (less than 10 per cent) of South Africa's *S. sagax* population.

This event, incredible to humans and sustaining for predators, is simply a long, shared voyage with a fatal end for the sardines.

Fish Scales

Much like chain mail, fish scales are the exterior armour of fish. They are designed to protect their owners from predators, which they achieve by having exceptional mechanical properties such as high strength- and toughness-to-weight ratios and resistance to damage. Additionally, fish scales are covered with a coating of mucus that helps ward off external parasites and fungi. These protectors have also evolved to facilitate locomotion, with highly mobile fish having special types of scales that accommodate the significant bending movement of their bodies and allow for faster movement by reducing drag.

Not all fish scales are created equal. They vary widely in size, shape, structure and extent. While some species, such as boxfish, have rigid armour plates made of scales, bottom dwellers such as eels and anglerfish are scale-less and depend on other forms of protection. Even sharks have scales, called placoid scales or denticles, which vary depending on how fast or slow they travel – faster-swimming sharks usually have overlapping scales, while the slower swimmers might have coarser, non-overlapping scales. Unlike bony fish, whose cycloid or ctenoid scales grow as they do and stay with them till they die, new placoid scales are added between older scales as the shark grows.

As the cycloid or ctenoid scales increase in size, growth rings called circuli become visible that resemble growth rings in a tree trunk. During cooler months, the scales grow more slowly and the circuli are closer together, leaving a band called an annulus. Biologists count these annuli to estimate the age of a fish.

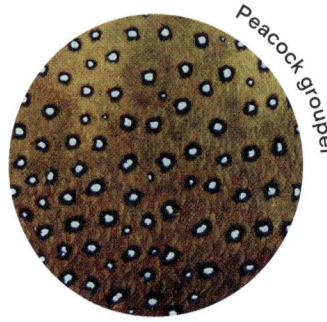

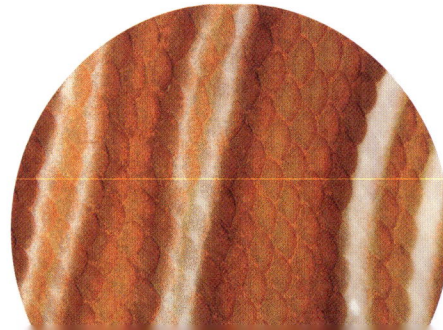

Peacock grouper

Juvenile emperor angelfish

Andaman damsel

Clown triggerfish

Fairy basslet

Lionfish

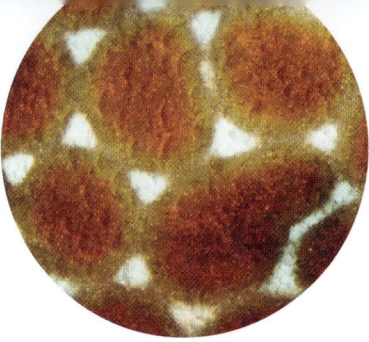

Grouper

Blue-lined triggerfish

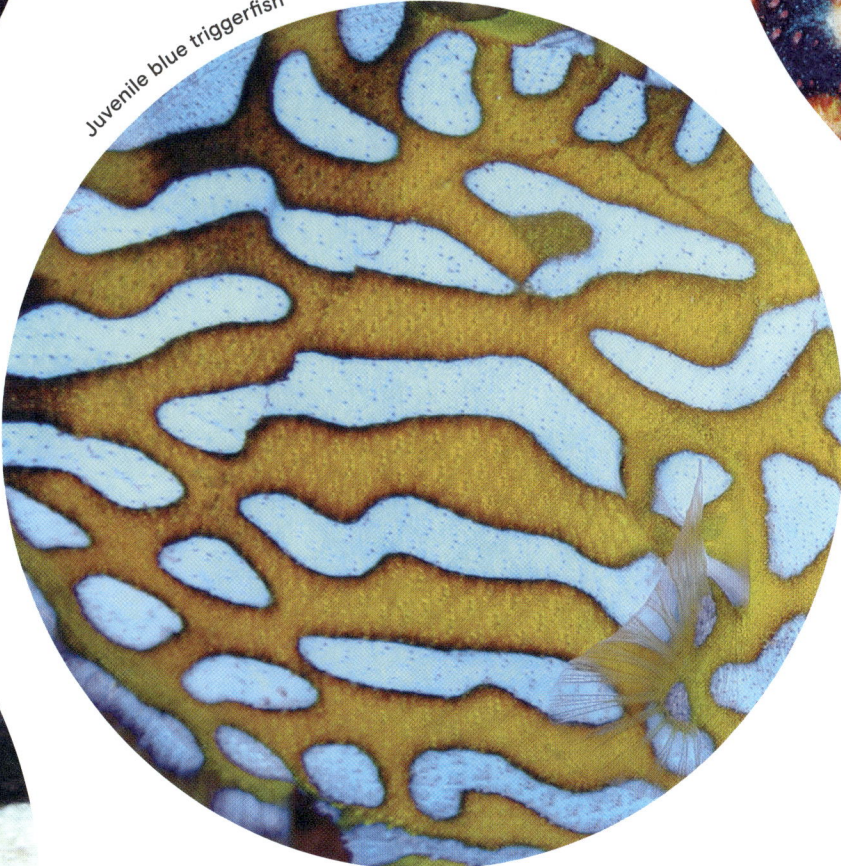

Juvenile blue triggerfish

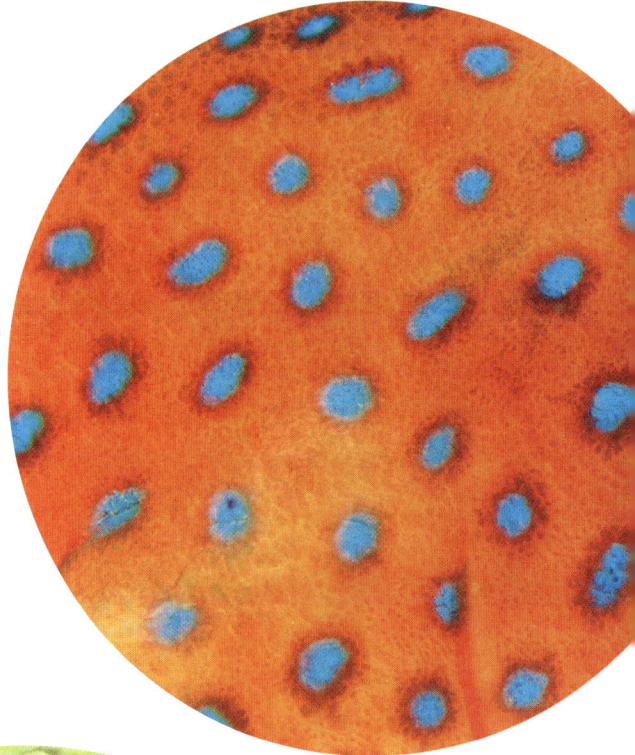

Coral cod

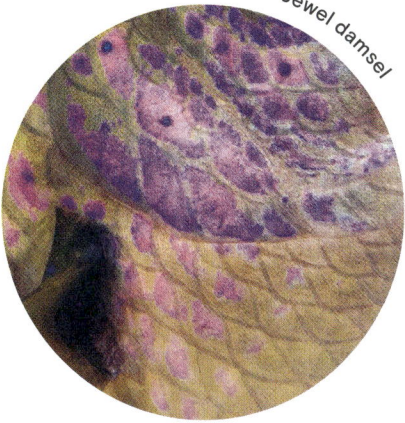

Jewel damsel

Butterflyfish

Extraordinary Divers

01

Northern gannets are the largest seabirds in the North Atlantic and boast a wingspan of up to 2 m (6½ ft).

Gannets search the oceans for tasty morsels from heights of more than 30 m (100 ft) using binocular vision, facilitated by forward-facing eyes that allow them to judge distance accurately while diving. Once prey is spotted, they waste no time folding their wings back and straightening their bodies to plunge into the water at speeds up to 100 km/h (60 mph). The dive can propel them up to 5 m (16 ft) underwater, but by using their wings and webbed feet, gannets can venture deeper still – up to 20 m (65 ft) below the surface – where they can remain for up to thirty seconds catching their meal. Thanks to their streamlined bodies, reinforced skulls, specially adapted neck muscles and a bone plate at the base of their bill, gannets can break the surface of the water without risking injury. Air sacs under the skin of the face and chest cushion them from the impact of the sea while also providing buoyancy when it is time to resurface. Meanwhile, the lack of external nostrils (nostrils are instead located inside the bill and covered by a flap of hard tissue when they dive) ensures they do not inhale any seawater on their way to lunch.

01 – Cape gannets
Morus capensis
Cape gannets diving for food

02 – Northern gannets
Morus bassanus
Making a grab for the same fish

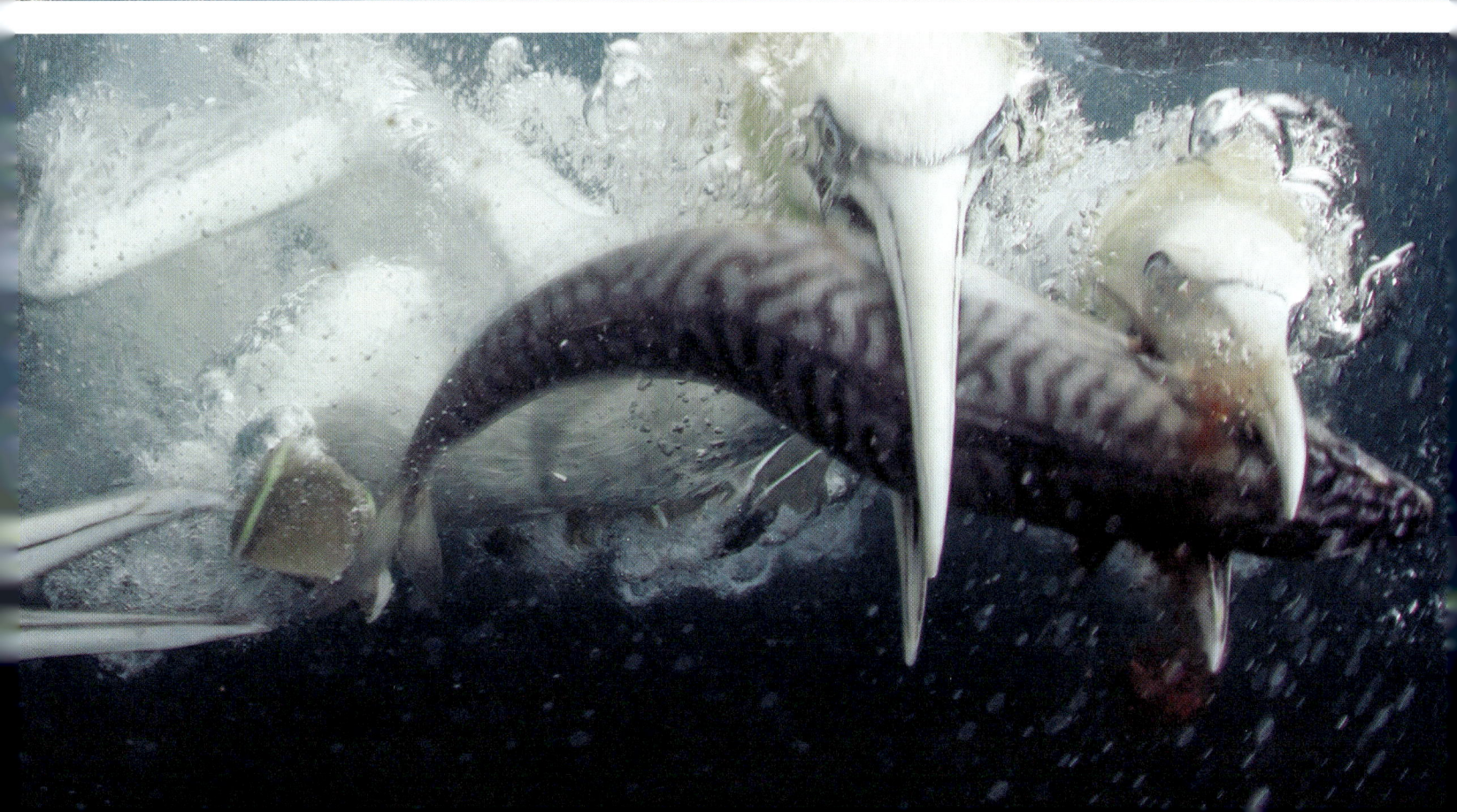

Tied to the Surface

Marine mammals are warm-blooded, have hair (at some point during their life), produce milk to nurse their young and breathe air through their lungs. Unlike fish, whose gills allow them to stay underwater throughout their lives, their dependence on the surface – regardless of the varying depths, temperatures, pressures, salinities and light levels they might experience over the day – means that the sunlight zone is important to their survival.

The depth to which marine mammals dive depends less on their size and more on their needs. Why dive deeper than your favourite food? Blue whales, the largest animals to inhabit our planet, are satisfied diving to depths of approximately 300 m (1,000 ft) for ten minutes, while sperm whales, the largest toothed whales, weighing in at just two-thirds to half the size of a blue whale, will dive deeper (approximately 2,250 m / 7,400 ft for one hour) in search of their prey. Cuvier's beaked whales, at their largest around half the size of a sperm whale, hold the record for the deepest divers in the ocean (almost 3,000 m /9,800 ft for over three hours). All three dive to depths that would kill humans, but sperm whales and Cuvier's beaked whales are aided by incredible adaptations such as high concentrations of the oxygen-carrying protein myoglobin in their muscles, and a highly mobile, collapsible ribcage that allows their lungs to be compressed during deep dives until they can return to the surface to replenish their oxygen and prepare for the next dive.

Depth in metres (feet)

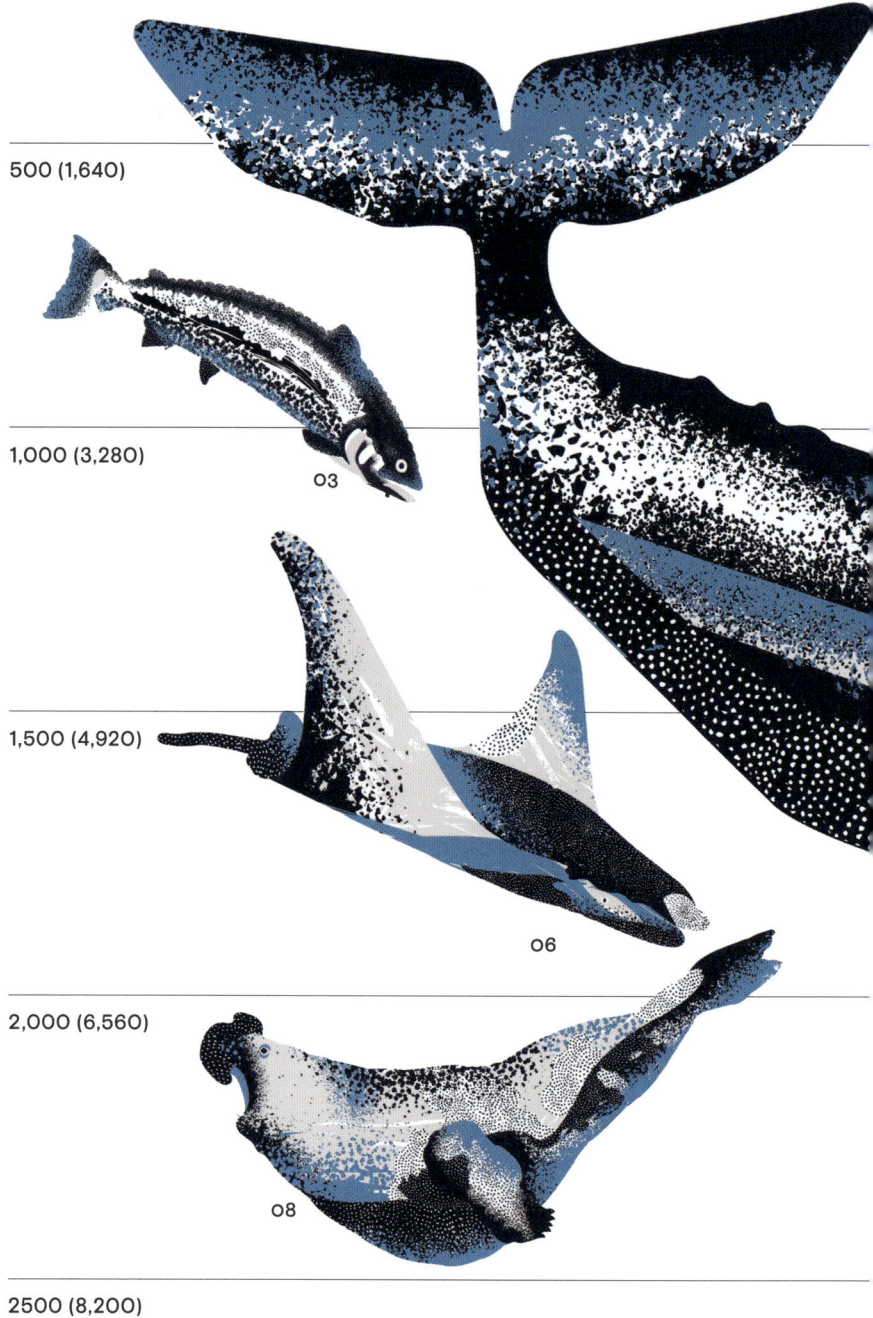

0

500 (1,640)

1,000 (3,280)

03

1,500 (4,920)

06

2,000 (6,560)

08

2500 (8,200)

3,000 (9,840)

01 – Emperor penguin *Aptenodytes forsteri* | 02 – Weddell seal *Leptonychotes weddellii* | 03 – Tuna Thunnini | 04 – Great white shark *Carcharodon carcharias* | 05 – Leatherback turtle *Dermochelys coriacea* | 06 – Chilean devil ray *Mobula tarapacana* | 07 – Sperm whale *Physeter macrocephalus* | 08 – Southern elephant seal *Mirounga leonina* | 09 – Cuvier's beaked whale *Ziphius cavirostris*

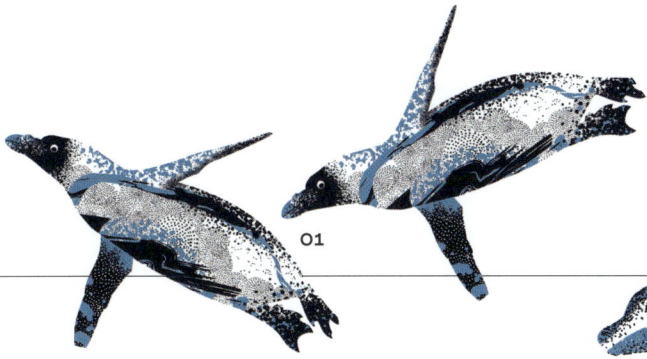

1

O1

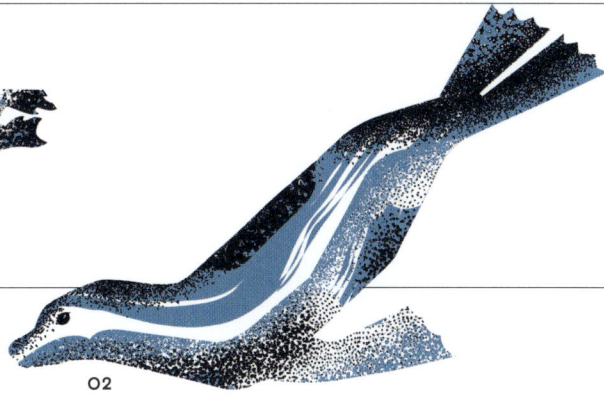

50

O2

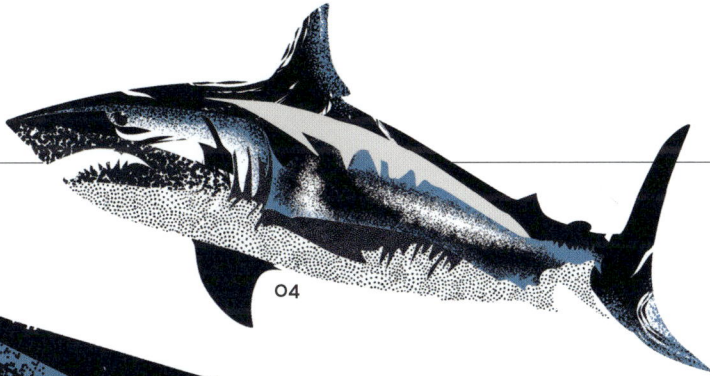

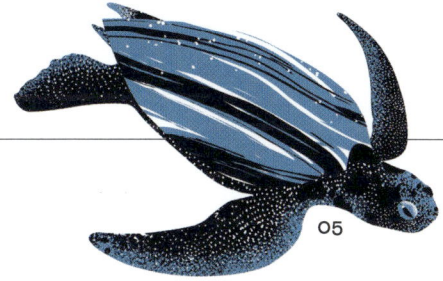

100

O4

O5

150

200

O7

250

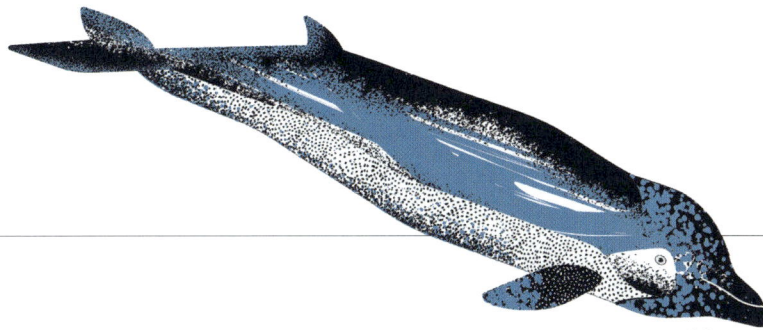

300

O9

THE SUNLIGHT ZONE

117

Banana nudibranch

Red-gilled nudibranch

Variable neon slug

Batangas nudibranch

Blue dorid nudibranch

Frosted nudibranch

Stealing Weapons for Self-defence

Nudibranchs are some of the most eye-catching creatures in our oceans. Despite not having a shell or any obvious protection, these soft-bodied sea slugs are proficient at defending themselves. Aeolid nudibranchs are, in fact, some of the greatest 'weapon-recyclers' in our oceans – they take weapons from their prey and use them for self-defence. They can do this thanks to their unique ability to feast on anemone and jellyfish tentacles packed full of little stinging cells called nematocysts – a feat not for the fainthearted. Adaptations such as chitin-lined mouths and throats mean they can bite and swallow tentacles without repercussions. The undigested nematocysts are then incorporated into the nudibranch's own cerata – fleshy, tentacle-like growths on their backs. If the nudibranch is threatened, the nematocysts are squeezed out of the cerata and discharged, either neutralizing or deterring the attacker.

Drummond's facelina

Black-margined nudibranch

Bullock's hypselodoris

Gender-fluid Coral Gardeners

Parrotfish, inhabitants of tropical and subtropical seas, are amazing in multiple ways.

First, parrotfish help to create our beaches. They do this by using their 'beaks' – actually large sets of fused teeth – to make sand. Scientists measured the hardness of parrotfish teeth near the biting surface and found it equivalent to the weight of a stack of about eighty-eight African elephants compressed to 6.4 sq. cm (1 sq. in) of space – or 530 tonnes. These fused teeth, or beaks, are made up of about one thousand teeth in fifteen rows, cemented together. Their natural hardness has led humans to use them as a prototype for designing ultra-durable synthetic materials, but for the fish, it means the ability to crunch coral all day long. Worn down teeth drop to the ocean floor and are quickly replaced by the next row of waiting teeth. These hardy teeth chew and grind the coral, which after ingesting, the parrotfish excretes as fine sand. A single parrotfish can produce a staggering 450 kg (990 lb) of sand a year.

As well as making sand, these fish also help to keep reefs healthy, highlighting their importance to marine ecosystems. Parrotfish clean the surface of coral by grazing on the algae that compete with coral polyps, allowing the coral to grow and become more resilient in the face of stressors.

Parrotfish are remarkable also because of their gender fluidity. All parrotfish (with the odd exception) start their lives as females. They form schools for the first few years of their lives until they are fully grown and sexually mature. At sexual maturity, the largest female undergoes a sex change to become what is known as a secondary male, who fertilizes the eggs of his old school during spawning aggregations. The sex change is reversible – these secondary males can turn back into females.

Before they go to sleep, parrotfish may do one other incredible thing. Using special glands behind their gills to secrete mucus, they produce a cocoon, much like a sleeping bag, in which they rest. The mucus cocoon takes between thirty and sixty minutes to create and is laced with antibiotics to keep away pathogens, while also protecting the parrotfish from parasites such as bloodsucking isopods (small crustaceans). In addition, it seals in the fish's body odour to ensure that hunters such as moray eels cannot sniff them out and take a bite.

The Ocean's Singers: Humpback Whales

All humpback whales use sound to communicate, but it is only the males that sing. That is not to say that females and calves do not communicate – they do – but their calls are relatively short and do not have a repetitive structure like song. When singing, the males hang vertically upside down, sometimes for hours. Songs are complex and haunting to listen to, and each population has a different song. Humpback whale songs are so famous that the 'Golden Record' launched into space in 1977 aboard the two *Voyager* spacecrafts as a kind of time capsule contained a recording of a humpback whale song collected by Dr Roger Payne in Bermuda in 1970.

To sing, humpback whales contract the muscles in their throat and chest, causing air from their lungs to pass across their vocal cords or U-fold within their larynx. The vibration of this U-fold is what produces the song. These songs are complex. They are made up of units – moans, cries and chirps – that are arranged into phrases. Lots of repeated phrases make themes, which are then sung in a pattern. Think of it as a poem where each phrase is like a line, and each theme is a stanza. When multiple themes (or stanzas) are arranged together in a specific sequence, that is a song (or poem).

Because males sing mostly during the mating season, their song may be used to attract a mate, or to warn off other males. While at the breeding grounds, all the males sing the same current rendition of a song, sometimes even singing together in choruses. Adult male humpbacks can sing in sessions that last from five minutes to more than twenty-four hours, repeating the same song over and over. While scientists initially believed that singing occurred only on the breeding grounds, it is now evident that they also sing when they feed. Why? No one is quite sure.

Over time, the songs evolve as one male adds a unit or changes a theme, which means that a song can differ from one population to the next. Fascinatingly, humpback whale songs spread from one population to another in the Pacific Ocean when the whales meet around their breeding grounds and learn a few new tunes. In just a couple of years, a song can spread thousands of kilometres.

While humpback whale songs are some of the best studied in the underwater world, they are also a gentle reminder that we still have so much to learn.

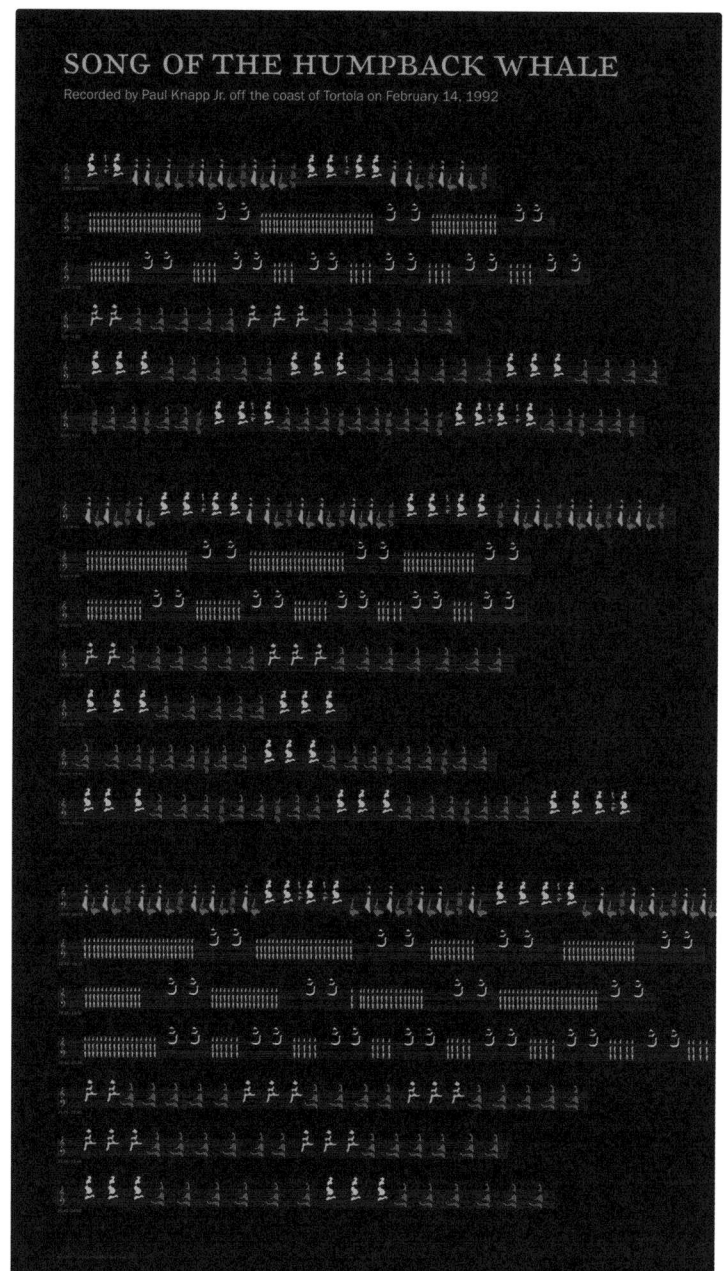

01

124

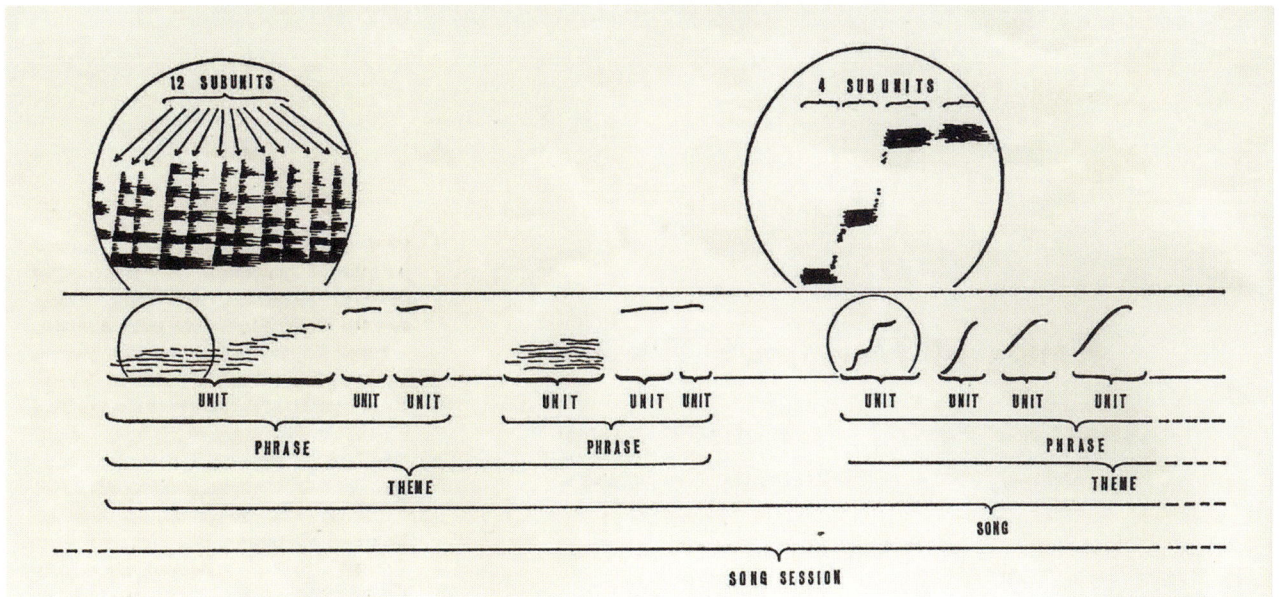

02

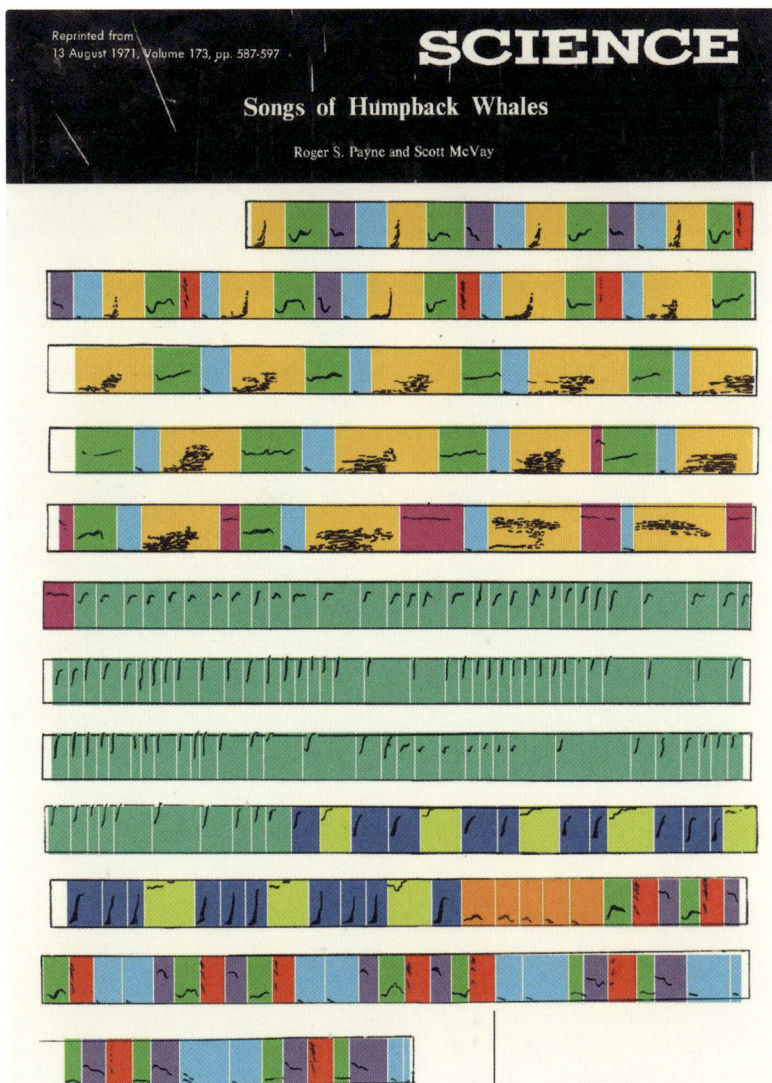

SCIENCE

Reprinted from
13 August 1971, Volume 173, pp. 587-597

Songs of Humpback Whales

Roger S. Payne and Scott McVay

03

04

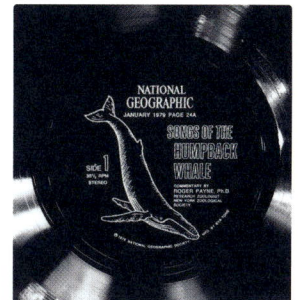

05

01 – Transcribed humpback
whale song

02 – An alternative notation
system for transcribing
whale-song patterns

03 – Another attempt to
colour-code and classify
units of the whale song

04 – *Songs of the Humpback
Whale* album cover

05 – *Songs of the Humpback
Whale* vinyl, 1979
Featured in *National
Geographic* magazine

Connectivity

The ocean is obviously important to species that spend their lives beneath the waves. But some species depend on it only at key stages, after which they make their way back to brackish or fresh water. These so-called 'euryhaline' species, characterized by their ability to tolerate a wide range of salinities, from brackish to marine, at some stage in their life cycle, include salmon. While salmon look different at each of the five stages of their life cycle, they also depend on different environments to support their growth and survival. This dependency on both marine and fresh water makes salmon an 'anadromous' species. Anadromous species switch between these environments, with one providing suitable conditions for reproduction while the other being better suited for feeding and growth. To survive, salmon switch their salt balance physiology, concentrating salts within their bodies in salt-deficient environments and excreting excess salts in saline seas or brackish water.

Salmon Life-cycle

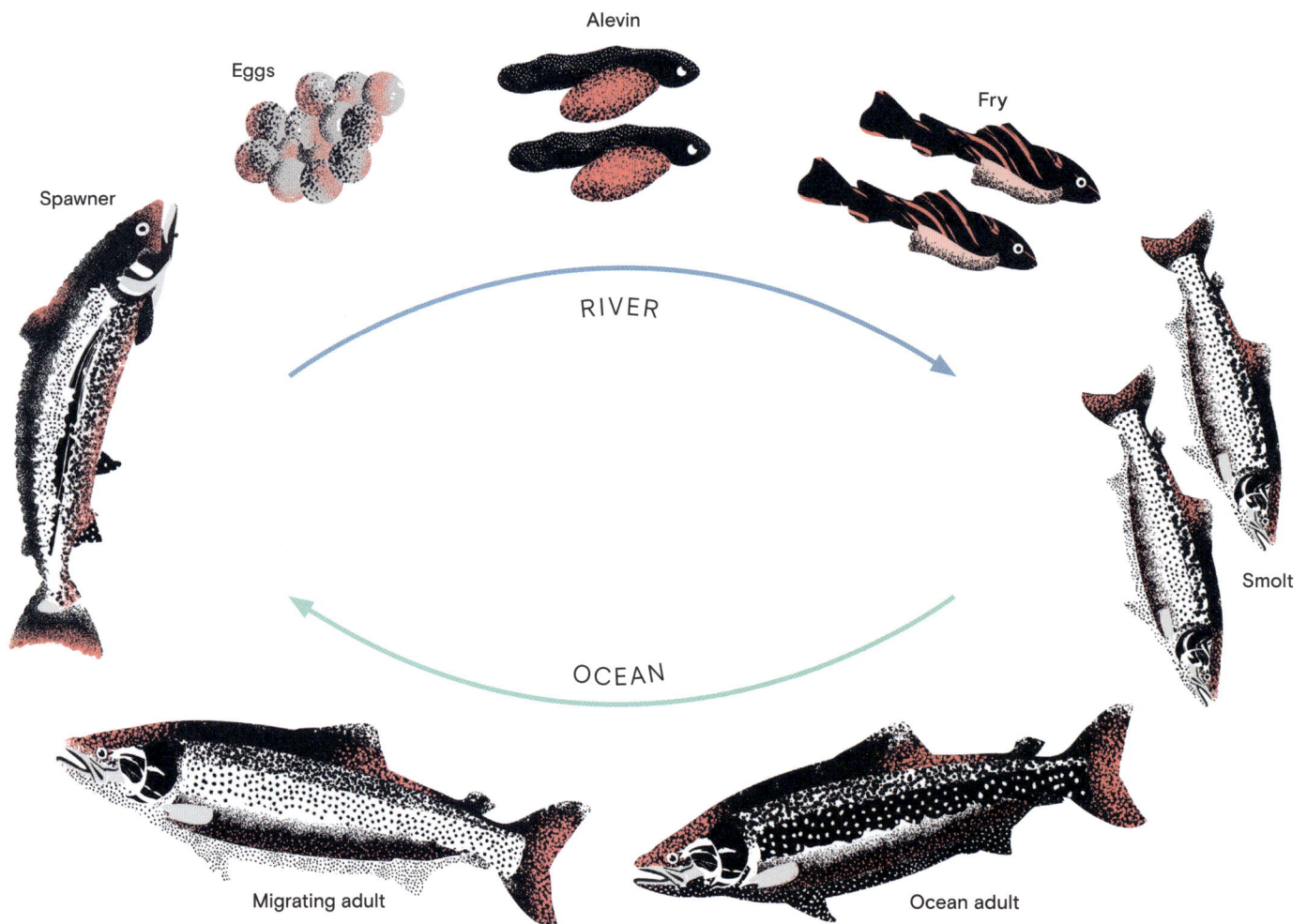

Eggs

Alevin

Fry

Spawner

RIVER

Smolt

Migrating adult

OCEAN

Ocean adult

Different Types of Salmon

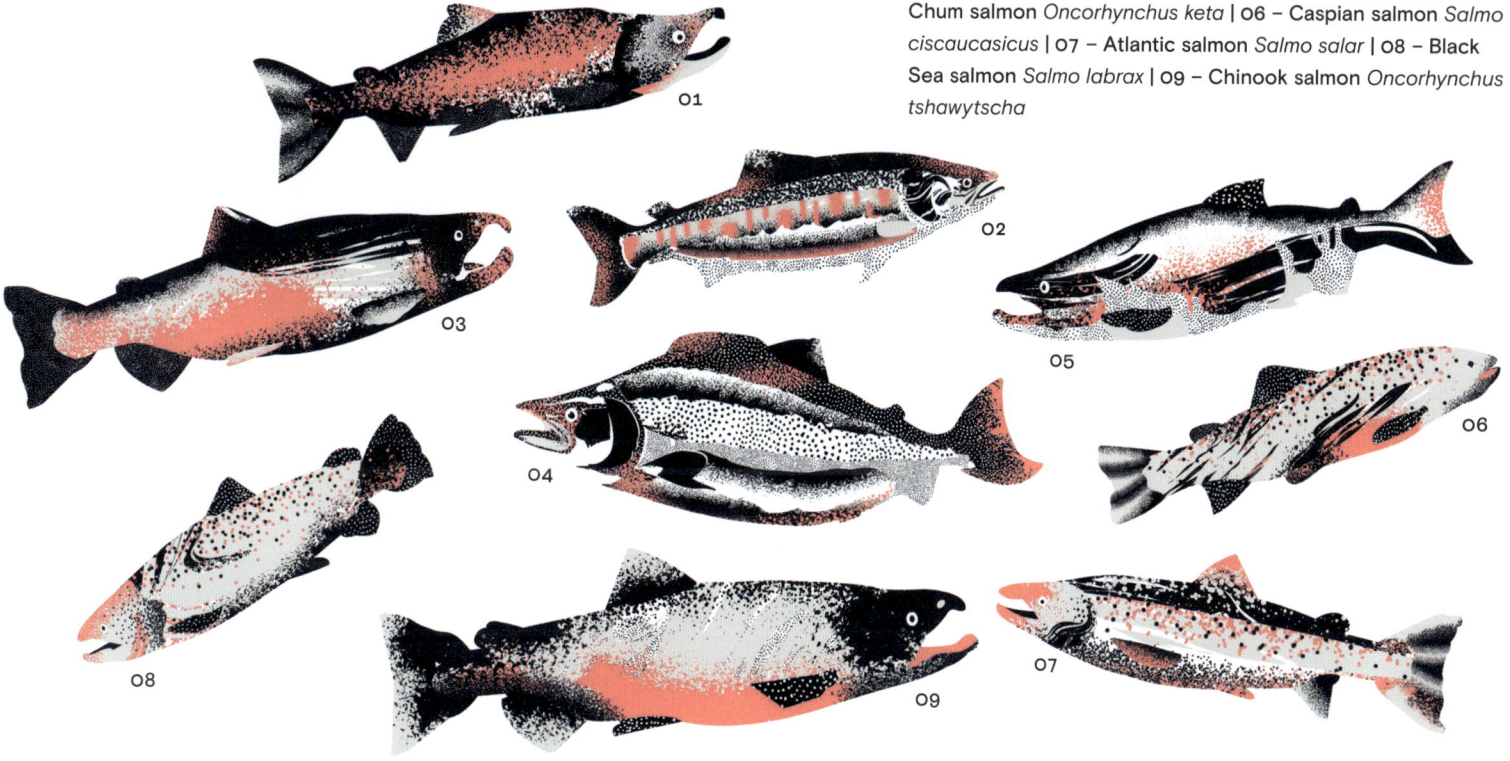

01

02

03

05

04

06

08

09

07

Osmoregulation in a freshwater environment

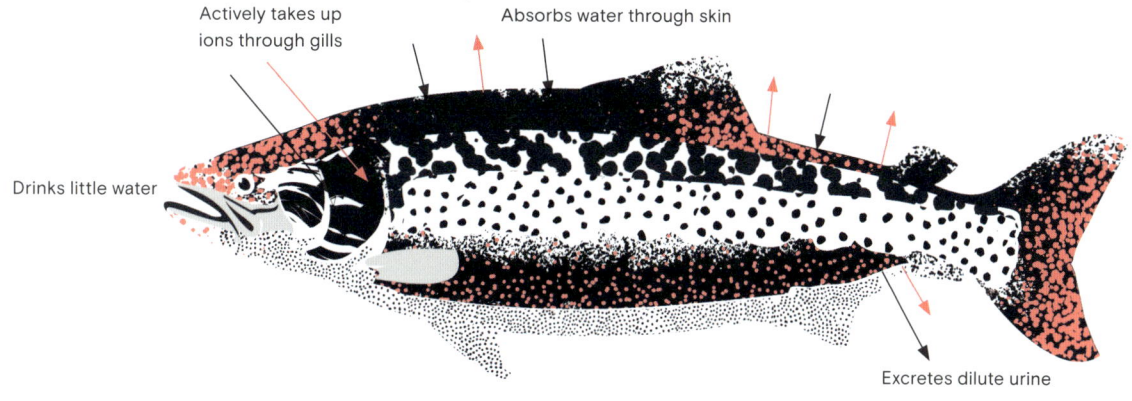

Actively takes up
ions through gills

Absorbs water through skin

Drinks little water

Excretes dilute urine

Osmoregulation in a saltwater environment

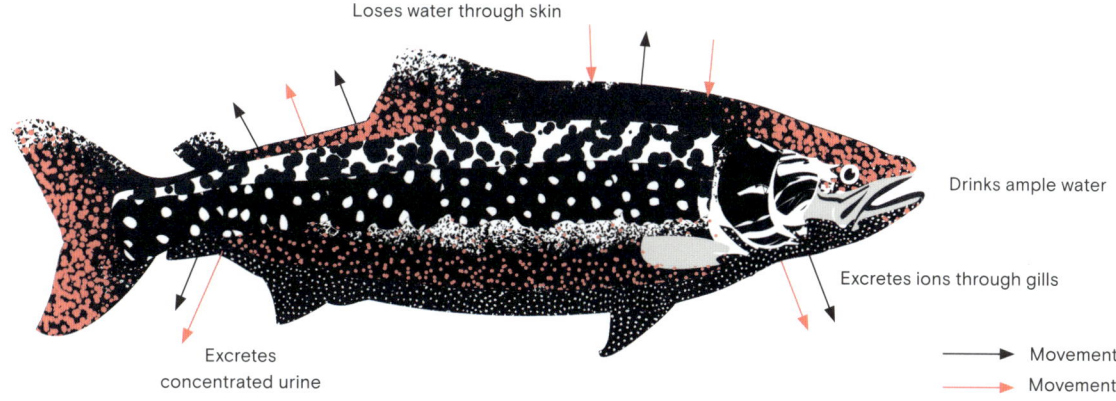

Loses water through skin

Drinks ample water

Excretes ions through gills

Excretes
concentrated urine

Movement of water

Movement of ions

Oriental sweetlips and cleaner wrasse

Orange-lined triggerfish and bluestreak cleaner wrasse

Green sea turtle and remoras

Cleaner wrasse swimming above a moon wrasse

Symbiosis in the Sea

Coral grouper with cleaner wrasse

Cleaner wrasse fish spend their entire lives 'cleaning', making them obligate cleaners. This might sound like a terrible deal, but in fact they get most of their nutrition by munching on parasites and mucus from their clients' skin, mouth and gills. This kind of close living arrangement is what is known as a symbiotic relationship. Symbiotic relationships always happen between different species, making them even more intriguing. In this instance, the cleaner wrasse, a species commonly found across the Indo-Pacific, will set up a cleaning station in a prominent place on the reef to serve other fish who need their parasites removed. The wrasse indicates that the station is open for business by swimming up and down, flashing its blue stripes, while the client indicates its interest in the service by coming to an almost complete halt with fins wide open. This show tells the cleaner it is safe to get on with the job.

The remora is another species engaged in symbiosis in our oceans. These fish, also known as suckerfish or sharksuckers, fasten to their hosts to get a free meal – feeding on their hosts' leftovers. It may look as though remoras attach to their hosts via their mouths, but in fact they have a unique suction disc structure, actually a modified dorsal fin, on the top of their flat heads, which sticks to the skin of the host. A fleshy ring of connective tissue around the periphery of the disc enables the remora to stay latched on without any extra energy expenditure. To release, it simply swims forward. While this may seem like an easy, almost lazy way of life, it is not. Remoras do not control where they travel – they follow their hosts to bathypelagic depths (between about 1,000 and 3,000 m, approximately 3,300–9,800 ft), where they are exposed to extreme gradients of light, dissolved oxygen, temperature and pressure; the adaptations that enable them to survive this journey remain a mystery. Remoras hitch rides on a range of large species, from the blue whale to manta rays, from whale sharks to turtles. They move around on their hosts removing external parasites. This form of symbiosis is called mutualism because both parties benefit – the remora rides free and feeds on the scraps of its host, while the host stays healthy and parasite-free.

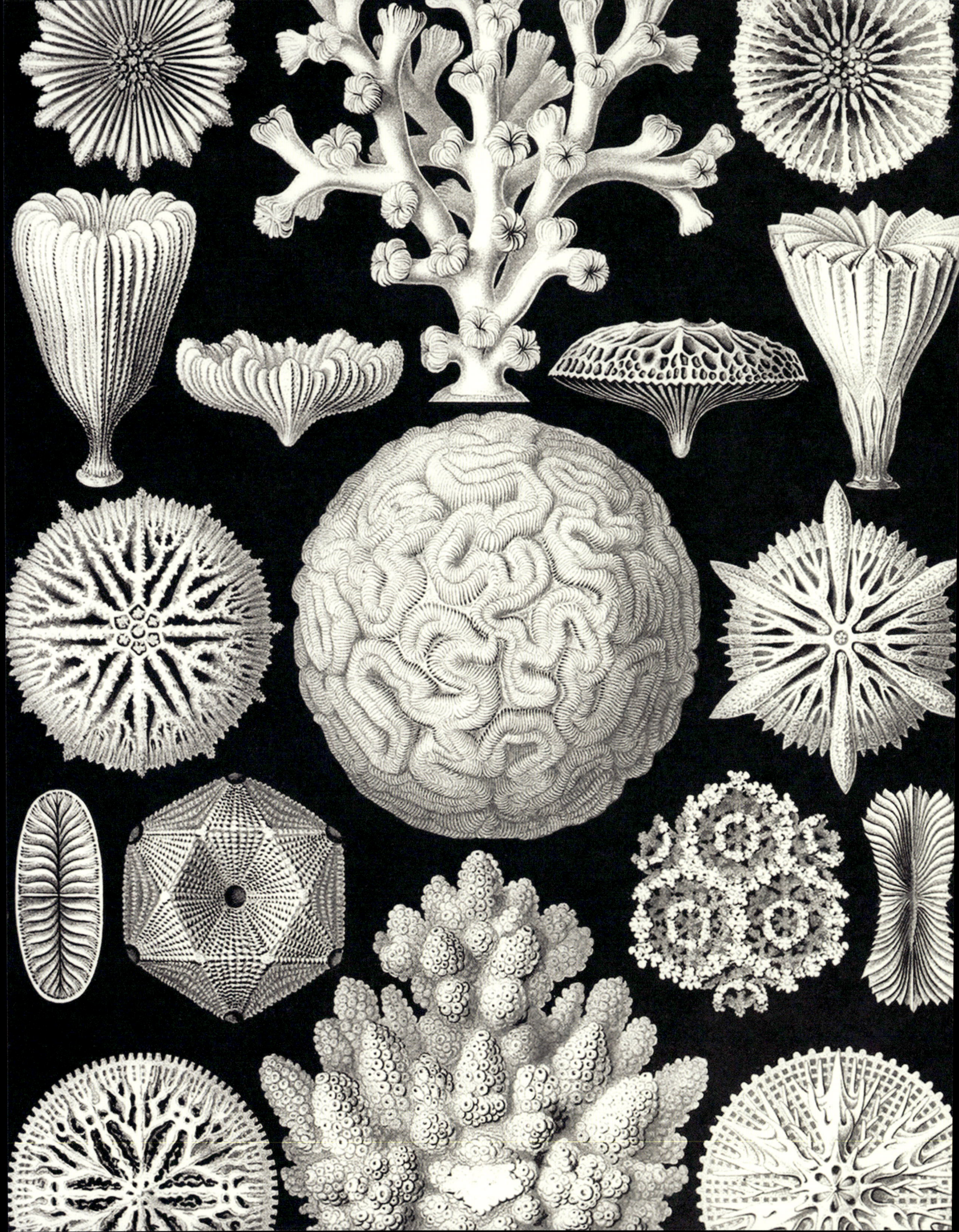

Ernst Haeckel: The Marriage of Art and Science

Before macrophotography – the magnification of the photographed subject – there was Ernst Haeckel. Haeckel (1834–1919) was a German biologist, naturalist, philosopher and physician. A field trip to the North Sea while at medical school sparked his lifelong fascination for natural forms and biology. Haeckel's ability to pay attention to detail, coupled with his meticulous art skills, helped him reveal the previously unseen world of microscopic organisms to the world (see pp. 162–63).

Best known for his work on radiolarians (minuscule protozoans), Haeckel was particularly captivated by the test – the elaborate mineral skeleton that surrounds the multiple body compartments of a radiolarian.

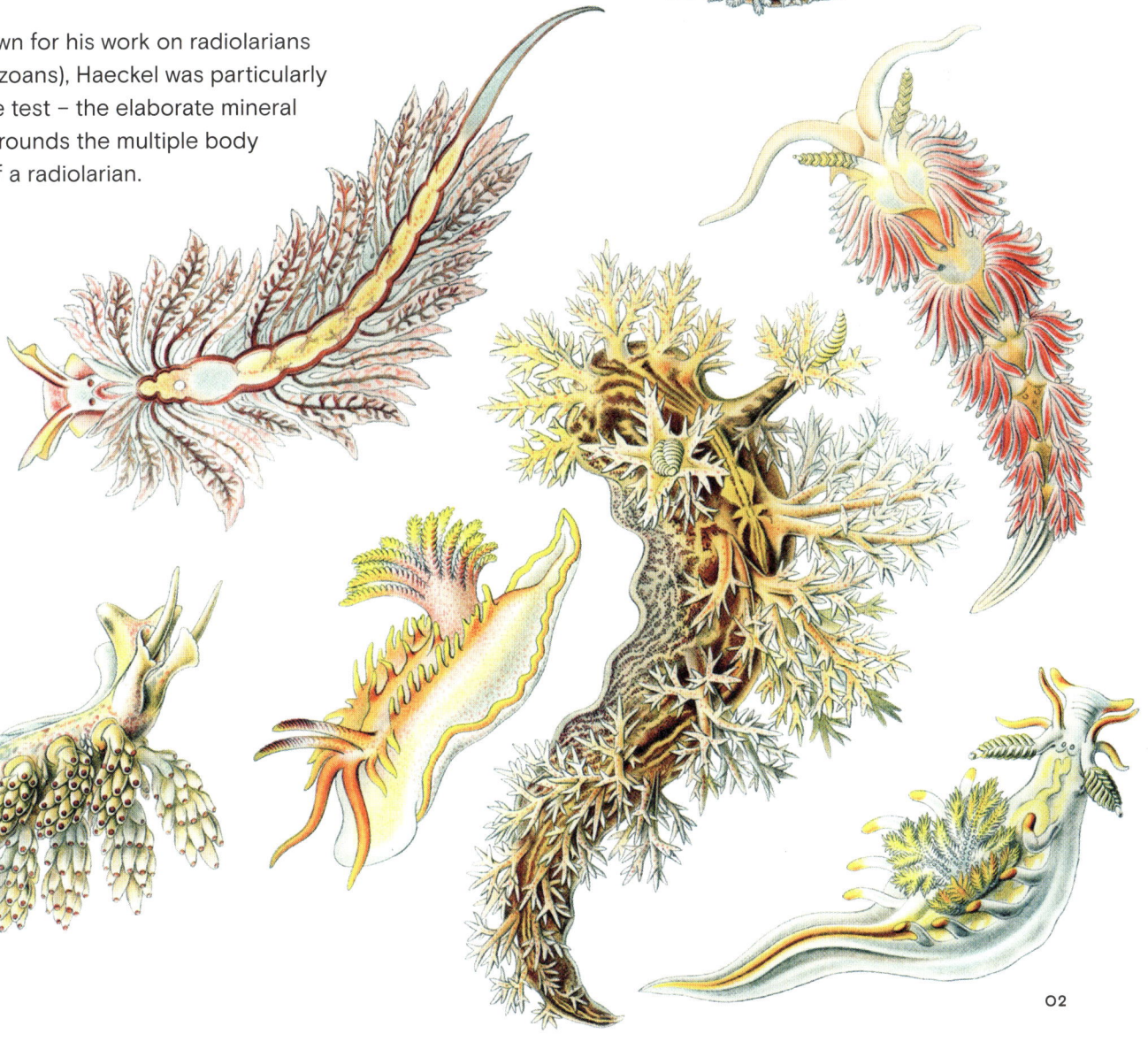

O2

Evolution of cetaceans

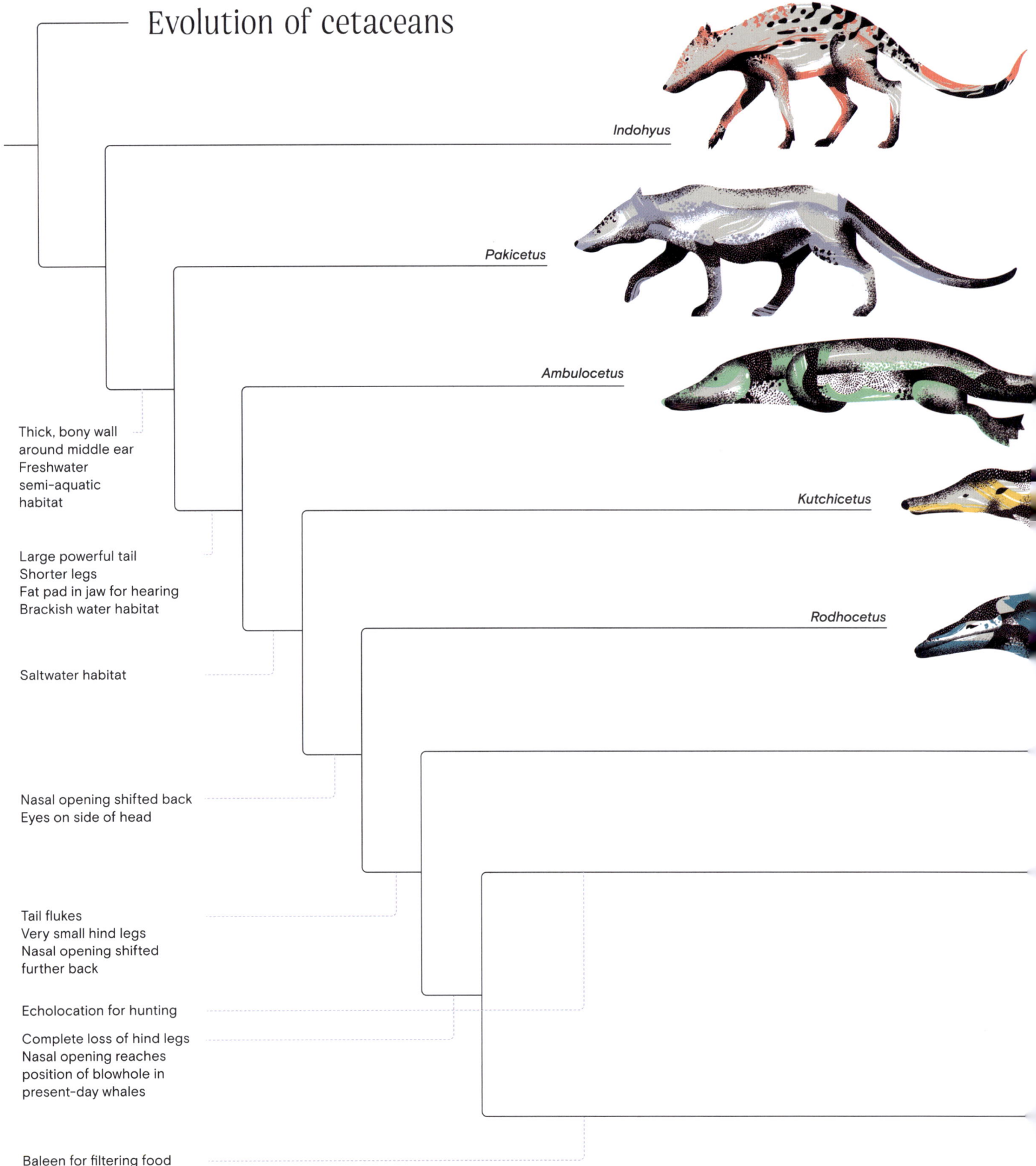

Indohyus

Pakicetus

Ambulocetus

Kutchicetus

Rodhocetus

Thick, bony wall
around middle ear
Freshwater
semi-aquatic
habitat

Large powerful tail
Shorter legs
Fat pad in jaw for hearing
Brackish water habitat

Saltwater habitat

Nasal opening shifted back
Eyes on side of head

Tail flukes
Very small hind legs
Nasal opening shifted
further back

Echolocation for hunting

Complete loss of hind legs
Nasal opening reaches
position of blowhole in
present-day whales

Baleen for filtering food

What's in a Tail?

Fish and cetaceans may live in the same habitat, but they have some essential differences. While fish are cold-blooded, use gills to extract air from water, and in most cases lay eggs, cetaceans are warm-blooded, breathe air through their lungs, and give birth to live young. Beyond this, there is one other key difference between the two groups – and this lies in the orientation of their tails. These tails, which are an extension of the vertebral column, provide clues to the evolutionary history of locomotion.

Fish evolved from ancestors that slithered along the seafloor, undulating from left to right to produce forward thrust. This resulted in vertically oriented tails. Conversely, cetaceans evolved from land-dwelling four-legged animals designed to run with their limbs underneath, and backbones that bent up and down to extend their spine and stride. In cetaceans, this resulted in horizontal tail flukes (fins). The vertical tail in fish means they can use muscles along the entire body to swim, while cetaceans use only the shorter muscles connected to the rear of the body, using their upper bodies for direction.

Dorudon

Odontocetes

Mysticetes

Evolution of Fish

Present Day

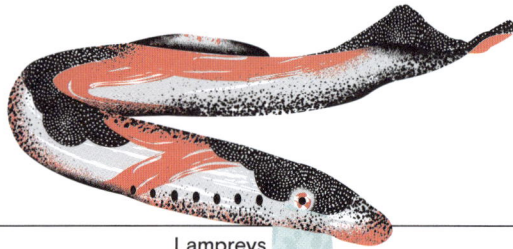

Sharks

Cenozoic

Lampreys

ELASMOBRANCHII

Cretaceous

Jurassic

AGNATHA

Triassic

Hybodus

Permian

Upper Carboniferous

Lower Carboniferous

Cladoselache

Devonian

ACANTHODII

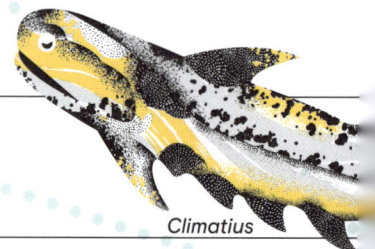

Silurian

Climatius

Ordovician

Hemicyclaspis

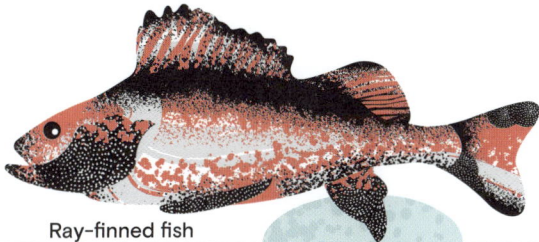

Ray-finned fish

Lungfish

TELEOSTEI

SARCOPTERYGII

Lepidotes

HOLOSTEI

CHONDROSTEI

Cheirolepis

Eusthenopteron

The Twilight & Midnight Zones

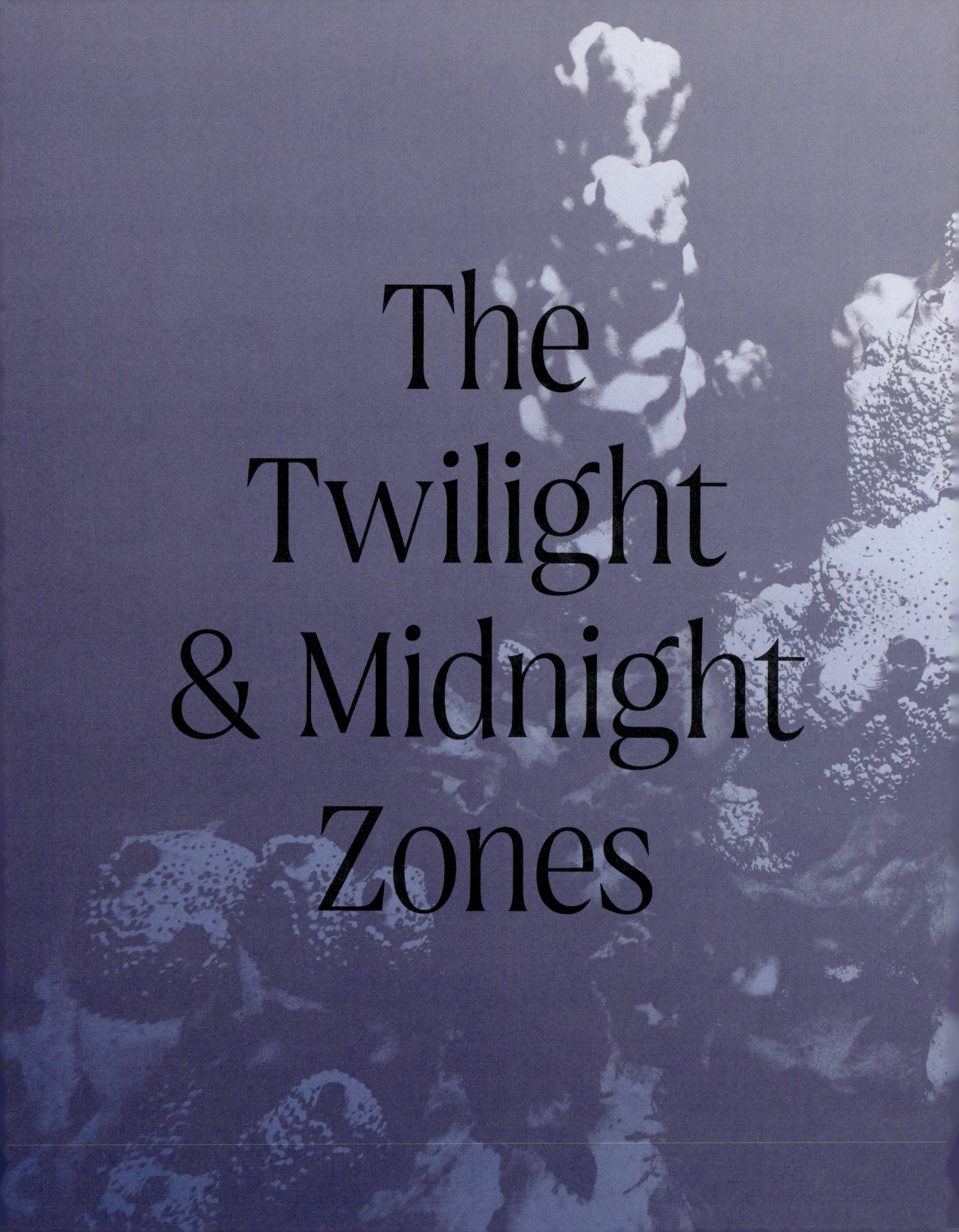

Introduction

Sinking beneath the sunlit zone, we leave the sun's warmth and radiance behind and plunge into the vast, mysterious depths of the ocean. It is easy to imagine the deep ocean as a single space that is uniformly dark and, certainly to humans, dismal and unwelcoming. A little over a century ago, leading scientists were convinced there was nothing alive past the first few hundred fathoms (a fathom is 6 ft or approx. 1.8 m). In their minds, the deep ocean was a huge lifeless void, too dark, too cold and too pressurized for anything to survive. Their assumption was far from the truth; the twilight and midnight zones are inhabited by an array of brilliant creatures, many of which literally light up the ocean.

Towards the end of the nineteenth century, a new wave of ocean scientists set out to find out what was really lurking down in the deeps. They stepped onto the decks of ocean-crossing ships – often converted navy vessels – equipped with research tools specifically designed to probe, measure and sample the ocean's furthest reaches. For months or years at a time these teams of scientists and sailors surveyed the global ocean, and as they carefully pulled in their nets and scooped up samples, they discovered that there is indeed life in the deep – all the way to the very bottom.

The species that deep-sea explorers have found are wildly different from anything living on land: there are glistening, beautiful creatures made from diaphanous strands of jelly; all manner of species that can light up their bodies and twinkle in the dark; monstrous-looking predators with gigantic teeth, inky-black skin and stretchy stomachs; twitching invertebrates with astonishing, elaborate eyes. The extreme conditions of the deep ocean have shaped these strange and extraordinary life forms. This is how life thrives in the dark, freezing, high-pressure depths.

Now that we know the deep ocean is full of life, it becomes clear that this is the single biggest portion of the planet's biosphere: more than 95 per cent of the space available for living things to occupy is comprised of the deep sea. It is the part of our world that scientists understand the least and still have so much to learn about.

A very few people have the chance to visit the deep sea themselves and peer out of the windows of submersibles capable of penetrating miles beneath the waves. More common are remotely operated deep-diving robots equipped with high-resolution cameras, which beam video in real time along cables stretching to scientists stationed on a ship at the surface, and sometimes transmitted live over the internet to anyone who wants to join in the remote deep-sea explorations. Watch the footage from these expeditions and sooner or later you will see something that will make your head spin in wonder. 'What the heck is that?' is a phrase uttered frequently by deep-sea scientists. It is anyone's guess how many species remain hiding and unknown down there, but judging by the unabated discovery rate there are many more to find.

Distinct layers in the deep ocean exist one on top of another, each with its own characteristic conditions and mix of species. Ocean scientists generally agree that the deep sea officially starts at the 200 m (650 ft) mark. This is the beginning of the twilight zone, or the mesopelagic, which is defined as the depth at which the sun is no longer strong enough to power photosynthesis. Of course, this is not a hard and fast depth – it is not like stepping over a line or drawing a curtain to shut out the light – and it is not precisely the same depth everywhere around the ocean or throughout the year. In cloudier, murkier waters, photosynthesis can grind to a halt at far shallower depths. However, even in the clearest waters, at 200 m (650 ft) all that is left of sunlight are the last remnants of dim blue light trickling down. Incidentally, this is why the ocean looks blue, because those short wavelengths of light plunge the deepest and scatter back from particles in the water.

The inky-blue light falling into the twilight zone cannot keep the sugar-making machinery working, and algae give up the ghost. Crucially this means that there is no new food being manufactured in the food webs

DEUTSCHE TIEFSEE-EXPEDITION 1898

The extreme conditions of the deep ocean have shaped these strange and extraordinary life forms. This is how life thrives in the dark, freezing, high-pressure depths.

A. BRAUER: TIEFSEEFISCHE I.   TAF. XV.

Taf. XV.
1-2 Melanocetus Krechi A. Brauer; 3 Melanocetus Johnsoni Günther; 4 Melanocetus vorax A. Brauer; 5 Melanocetus pelagicus A. Brauer;
6 Oneirodes niger A. Brauer; 7 Ceratias Caueri Gillz; 8-9 Gigantactis Vanhöffeni A. Brauer.
Verlag von Gustav Fischer in Jena.

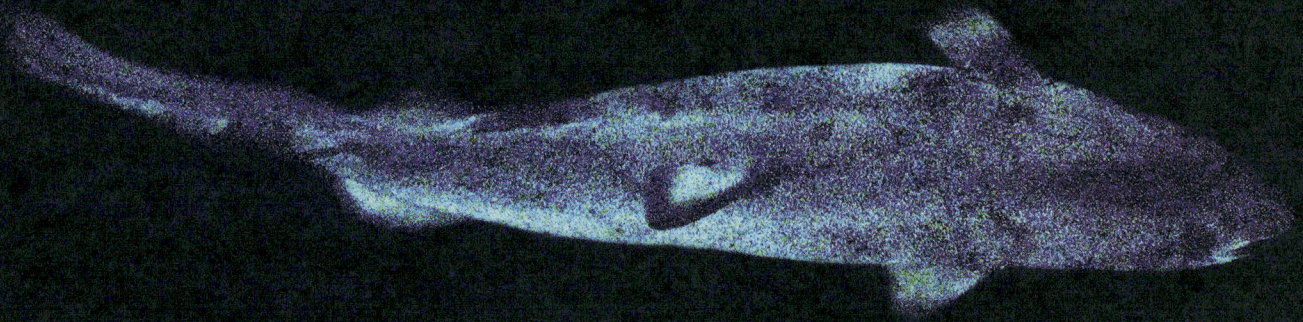

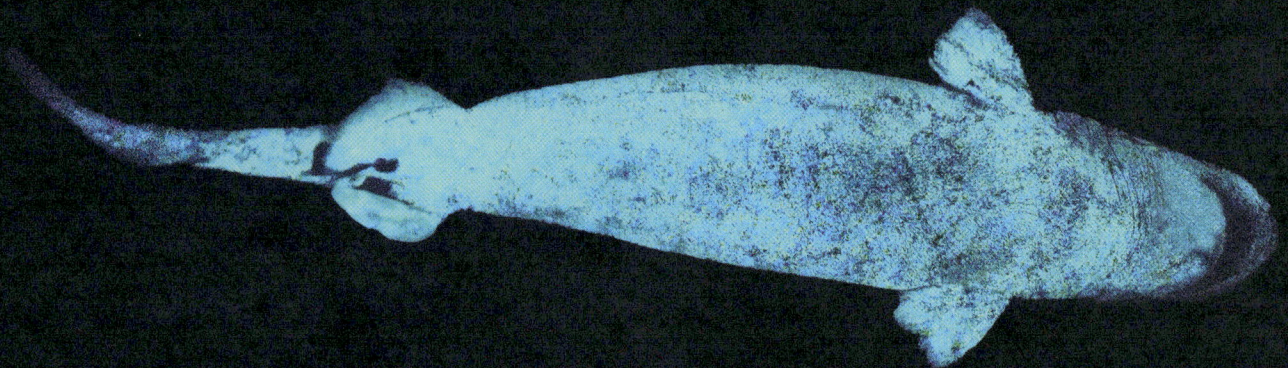

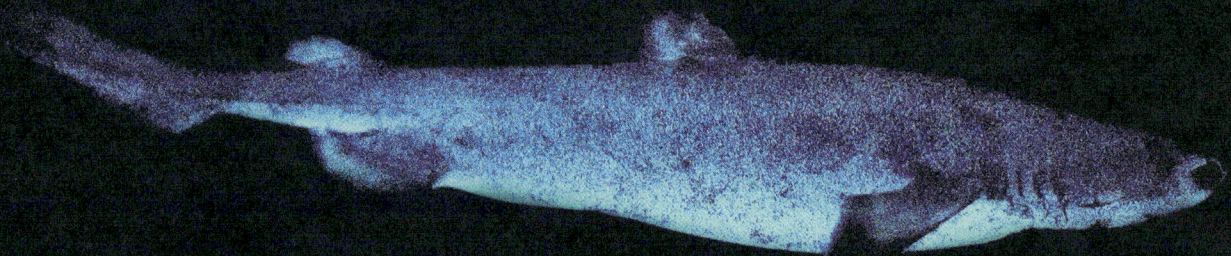

01 – Kitefin shark
Dalatias licha
Three sides of the kitefin shark,
the only bioluminescent shark

02 – Hairy sponge squat lobster
Lauriea siagiani
Raja Ampat, West Papua,
Indonesia

03 – Hermit crab larva

of the twilight zone or lower (with a few exceptions, as we will see, deeper down). This limitation on food production dominates life at great depths. To survive in the deep ocean, animals must be either predators, chasing after other animals, or scavengers, waiting for dead, decomposing remains to drop down. Among the immense diversity of life in the twilight zone, predators and scavengers have evolved many ways to make do and thrive where there is little food to be found.

To a human inside a submersible descending into the twilight zone, it soon appears completely dark outside the window. However, many animals living here have evolved sophisticated, super-sensitive eyes that can detect the last dim rays of light. This means that it is critical for animals to find ways to conceal themselves in these huge open waters, where there is nothing to hide behind.

The twilight zone continues down to 1,000 m (3,300 ft), where it gives way to the next layer of the deep ocean: the midnight zone, or bathypelagic. This realm lies completely beyond the reach of sunlight. Day or night, it is permanently dark in the midnight zone, except for the flashes and sparks of light made by the animals that live there.

Inhabitants of the midnight zone face the same challenges as those in the twilight zone, only more so. Pressure keeps ramping up and food becomes ever scarcer and harder to come by. But still, life is here in a dazzling array of strange and wonderful forms.

The midnight zone is an immense layer, 3,000 m (nearly 10,000 ft) thick, stretching to the 4,000 m mark, or close to 2½ miles underwater. This is roughly the average depth of the global ocean, but in many places this is not as deep as it gets, and below the midnight zone there is still more to come.

02

03

THE TWILIGHT & MIDNIGHT ZONES: INTRODUCTION

Hunting Giant Squid

People have long known that far beneath the waves, sperm whales hunt for squid. From the seventeenth century onwards, commercial whalers from Europe and North America killed hundreds of thousands of sperm whales for the valuable oil inside their enormous heads, and often found their stomachs filled with the indigestible mouthparts of squid. Judging from their gut contents, sperm whales feasted on medium-sized squid, a metre or two (3–6 ft) in length, and sometimes ate the biggest species of all – giant and colossal squid – which can be as long as a bus from tail to tentacles. Large circular scars on sperm whale skin were evidence of the battles that raged in the depths, as squid grappled their attackers with sharp hooks on their sucker-clad arms.

Much more recently, people learned how whales hunt for squid. With the help of electronic motion detectors and hydrophones fixed to these deep-diving giants, researchers discovered that sperm whales regularly plunge at least 1,000 m (3,300 ft) down and can hold their breath for an hour or longer. And, like bats, sperm whales use beams of sound to hunt in the darkness of the twilight zone.

Sounds emanate from snorts of air pushed along a sperm whale's right nostril, which vibrate flaps, called the monkey lips, that act like a human voice box. The hunt begins as a whale fires volleys of clicks into the water, one or two per second, scanning the surroundings. When it hears echoes bouncing off something promising, the whale speeds up the clicks until it sounds like a creaky door. As the whale nears its prey, it stops beating its tail and glides up silently, then puts on a sudden burst of speed, firing more shots of sound as the squid tries to escape. When it comes within striking range the hunter performs a handbrake turn, stopping suddenly and sucking in the squid whole without chewing.

Sperm whales have a huge appetite for squid, eating between 100 and 500 every day, which weigh around a tonne in total. To fill their bellies, adult sperm whales spend roughly three-quarters of their lives in the twilight zone, making only brief visits to the surface to replenish their oxygen supplies before plunging back down into the dark.

SPERM WHALES AT DINNER.

[To face p. 20.

01

01 – A pod of sperm whales hunting for squid, illustrated in *Creatures of the Sea* by Frank Thomas Bullen (1904)

02 – A sperm whale and giant squid in battle, illustrated in *Creatures of the Sea* by Frank Thomas Bullen (1904)

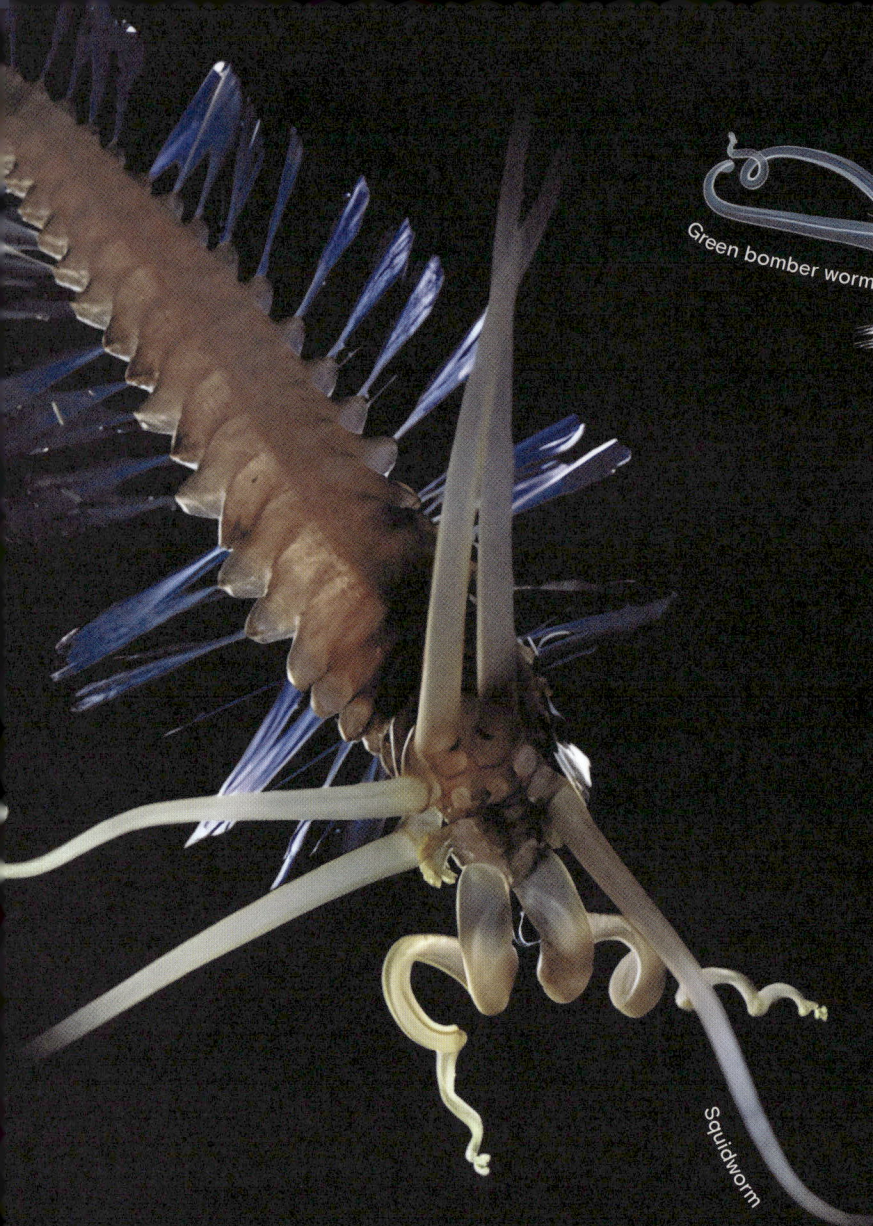

Green bomber worm

Squidworm

Gossamer worm

Bristle worm

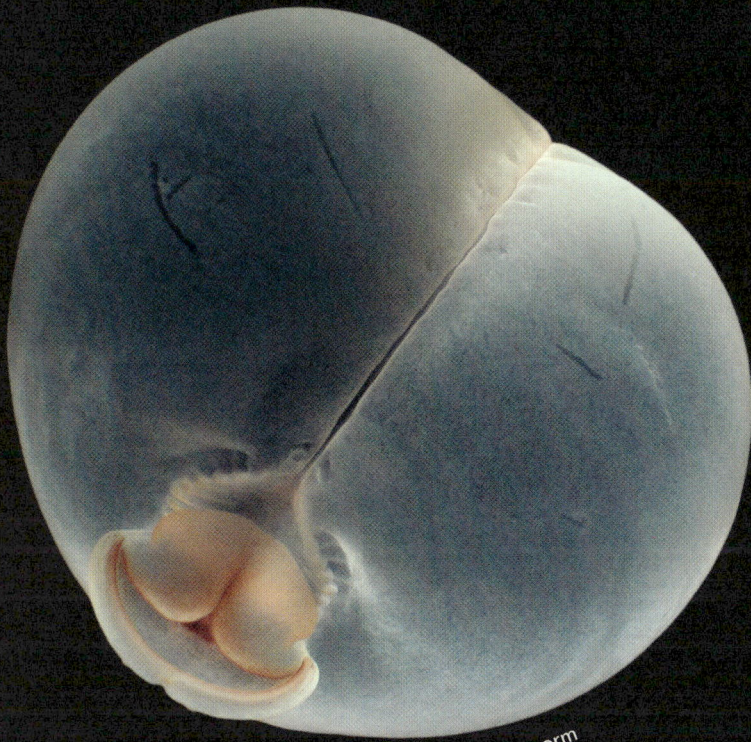

Pigbutt worm

Swimming Worms

In the twilight and midnight zones, worms do not just crawl across the seabed and burrow through sediments as their earthbound cousins do. Instead, many species have evolved to swim off and occupy the huge open space of midwater.

Gossamer worms look like glistening centipedes that twirl elegantly through the water, swimming forwards and backwards with equal ease, rippling their paddle-like appendages in sinuous waves. Squid worms have long tentacles, like spiralling tendrils of blown glass, which sense prey and predators nearby. When they are scared, green bomber worms hurl glowing orbs to startle an attacker and allow them to make their escape into the dark.

The aptly named balloon worms and pigbutt worms take a less active approach to life. Part of their bodies have inflated into a fluid-filled bubble, which lets them effortlessly float in the water column, conserving their energy reserves.

Balloon worm

Gossamer worm

Being Ultra-black

Many species of dark-coloured fish swim through the deep sea, including some that are not just black, but ultra-black. Among the very blackest fish are dragonfish, viperfish and lanternfish, dreamers, star eaters and pelican eels, which all look like they've been dipped in liquid soot. Taking detailed photographs of them is just about impossible even with the most sophisticated lighting equipment, because their skin absorbs almost all of that light. Their portraits usually come out as featureless silhouettes.

Scientists have studied sixteen species of ultra-black fish that absorb more than 99.5 per cent of the light that falls on their skin. To put that in context, a sheet of black construction paper absorbs around 90 per cent of the light falling on it, and a car tyre absorbs around 99 per cent. The blackest fish are as black as the artificial material Vantablack, which is constructed from carbon nanotubes. Fish make themselves ultra-black in a very different way. Like other vertebrates, including humans, fish use the pigment melanin to absorb light in their skin. This pigment is packed into tiny cellular compartments called melanosomes, which in deep-sea fish are the ideal size and shape to scatter light sideways. Any photons (light particles) that are not immediately absorbed by a melanosome are directed towards its neighbours, which then suck up any remaining glimmers of light.

The discovery of the superefficient light trap in deep-sea fish skin has got engineers excited because it has shown them a new way of making ultra-black materials, which are used in optical technologies such as telescopes and cameras. The way fish do it could lead to much cheaper, more flexible and more durable constructions.

Fish are not the only ultra-black animals. Butterfly wings and bird feathers that absorb a lot of light have also evolved, although these animals use their blackness to offset their vibrant colours and put on eye-catching displays to woo potential mates. For deep-sea fish, being ultra-black is all about not being seen in the reflected light of other glowing animals, and for some it is about not being seen for who they really are. Many fish combine their black skin with bioluminescent lures on chin barbels and forehead prongs to tempt prey towards them. Their deception would not work if the lure illuminated the hungry fish lurking behind.

01 – Two species of 'lantern-bearing seadevils' from *The Arcturus Adventure* (1926)

02 – Several species of bristlemouths (*Cyclothone* spp.) illustrated in the report of the *Valdivia* Expedition of 1898–99

Dragonfish

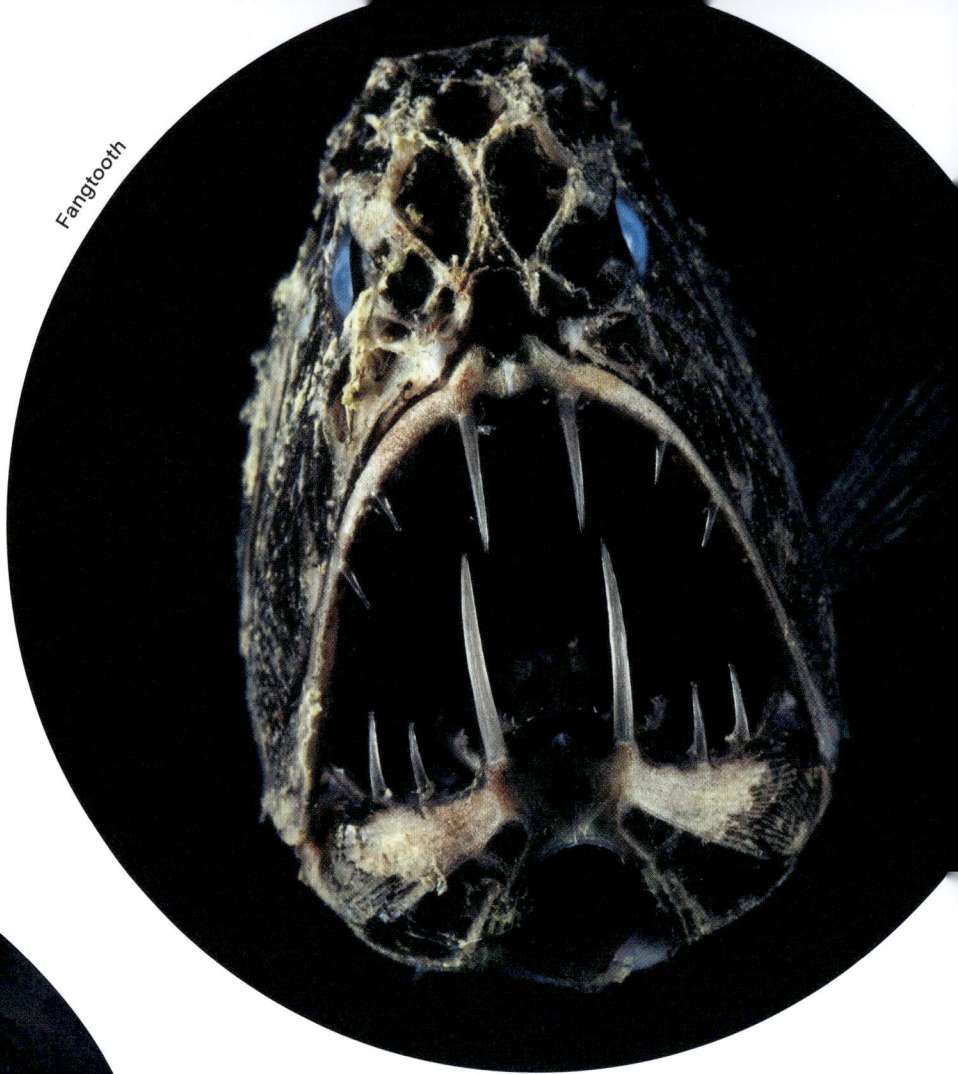

Fangtooth

Black swallower

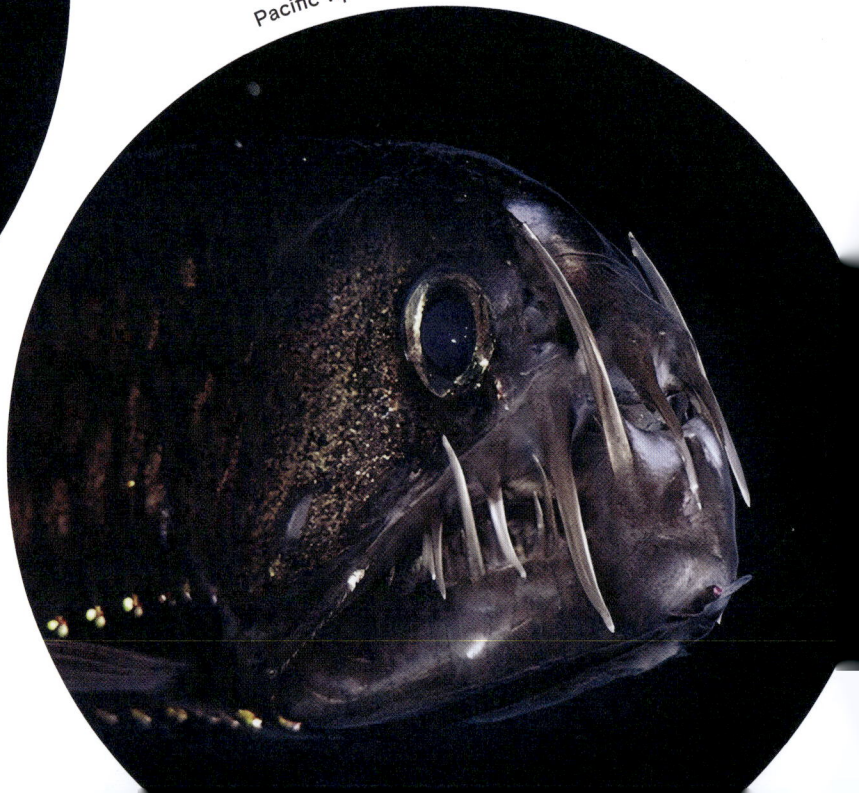

Pacific viperfish

The Hunt

Black seadevii

Compared to the food-filled, sunny shallows of the ocean, the twilight and midnight zones are far hungrier places, where there are not as many opportunities for finding nourishment. For the most part, deep-sea animals either rely on catching and scavenging what falls from above, from flakes of marine snow to the giant bodies of dead whales, or they eat each other. And in the deep sea, predators cannot afford to be fussy. That is why among the hunters of the twilight and midnight zones we see spine-chilling teeth, cavernous jaws and gigantic stomachs – all so they can catch and gulp pretty much anything that comes along, no matter its size.

A common strategy is sit-and-wait predation, also known as ambush predation. Hunters float in the dark as inconspicuously as possible (many have ultra-black skin to help them hide; see pp. 150–51), and either tempt prey to come and take a closer look with a glowing lure dangling from their body, or just wait for something to come within range. Then, in the blink of an eye, there is a snap of jaws and teeth, and the hapless target becomes dinner.

Gulper or pelican eel

Various deep-sea anglerfish species have vicious-looking mouths lined in needle-like teeth that point backwards, helping to prevent any wriggling prey from escaping. Pelican eels, also known as gulper eels, have jaws that take up roughly half their body length and unfold like an umbrella to engulf their prey. Viperfish have glassy teeth that are so long they do not fit inside their mouths, and instead interlock and create a deadly cage to trap prey.

A common misconception is that these are enormous predators. In fact, many deep-sea hunters are relatively small and would fit in the palm of your hand. Most anglerfish are football-sized or smaller. Still, they can fit a lot inside their bodies with the help of stretchy skin and a distensible stomach. X-rays and CT scans of preserved museum specimens show deep-sea fish with ambitiously large prey crammed inside.

Deep-sea lanternfish

Deep-sea Food Webs

Many fragile, diaphanous animals spend their lives wafting through the open waters of the twilight and midnight zones. Jellyfish and their various relatives, such as siphonophores (see pp. 162–63), consist largely of jelly, which is basically a thin mix of water and the protein collagen. It is an efficient, low-cost way of building a body, ideal for life in the hungry deep sea, with the added bonus of being naturally buoyant – these gelatinous creatures need not expend too much energy swimming. They are undoubtedly delicate creatures. If you tried to pick one up it would likely melt right through your fingers. But that does not mean these animals are insignificant in their realm.

An archive of underwater footage shot over thirty years by researchers at the Monterey Bay Aquarium Research Institute (MBARI) in California has revealed how important jelly-based animals are in deep-sea food webs – much to the surprise of scientists, who had rather assumed that not much eats these wobbly bags of water. Remotely operated underwater vehicles, roughly the size of a small car, are sent miles beneath the waves equipped with high-definition cameras to survey and record what is going on down there. Scouring MBARI's archive, researchers picked out every instance where deep-sea animals were caught on camera in the act of eating or being eaten. In hundreds of images, they saw that jellyfish and their like are prolific both as prey and as predators, even though they do not fit the standard model of deep-sea hunters with sharp teeth and huge jaws. For instance, dinner plate jellies (*Solmissus*), which do indeed grow to around the size of a dinner plate, have been spotted catching dozens of different prey types. Jellies eat worms, krill and other jellies. Fish eat jellies, as do squid and octopuses.

By untangling deep-sea food webs, it is becoming clear that gelatinous animals are not simply floating about, catching the occasional flake of marine snow here and there, then falling to the seabed when they die. They play critical roles in the ecosystems, linking up parts of the deep ocean in ways we are only just beginning to understand.

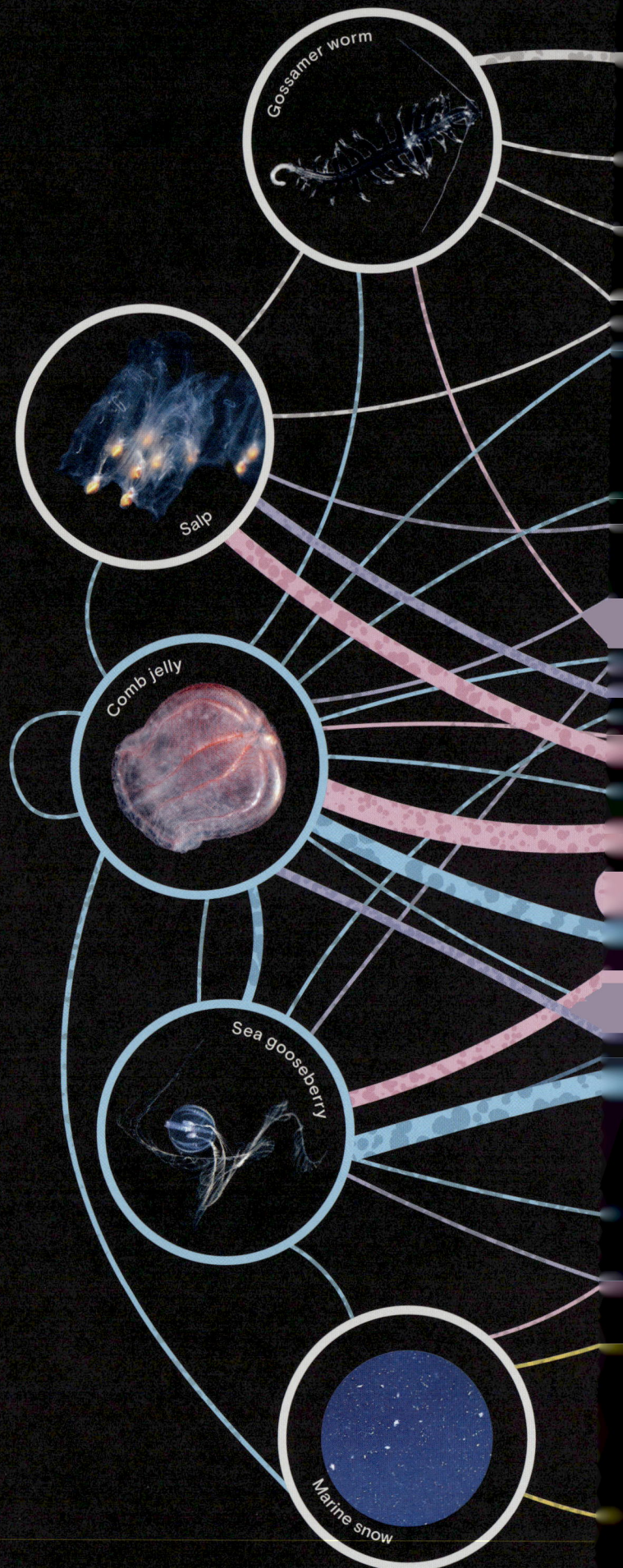

Arrow worm

Scyphozoa jelly

Cephalopod

Siphonophore

Deep-sea medusa

Deep-sea fish

Physonect siphonophore

Narcomedusa

Deep-sea shrimp

Trachymedusa

Deep-sea copepod

Radiolarian

Deep-sea isopod

Calycophora siphonophore

Sea butterfly

Snow in the Deep

In the 1950s, two Japanese scientists, Noboru Susuki and Kenji Kato, looked out of the window of an undersea observation chamber and watched flurries of fluffy white particles drifting past them and sinking into the deep. They later named these particles marine snow and put forward the idea that they made an important link between the ocean and living matter within it. And they were right: many animals do indeed eat marine snow.

The source of marine snow is far less delightful than the name suggests. It is largely made of dead phytoplankton, as well as zooplankton and their droppings, all stuck together with microbial slime. Nevertheless, it is incredibly important stuff. This rain of organic material is the main form of food that reaches the base of deep-sea food webs.

The amount of marine snow that falls varies throughout the year and from place to place across the ocean. When spring blooms of phytoplankton turn the surface seas into swirls of green and turquoise, they are followed by great seasonal pulses of marine snow. All in all, however, only around 2 per cent of the food produced at the sea surface will end up sinking and landing on the deep seabed, where sea cucumbers, starfish and other scavengers roam the abyss searching for freshly fallen piles of snow.

On the way down, lots of animals are doing their best to catch the snow as it falls. Tiny swimming snails, known as sea butterflies (or pteropods), cast sticky nets into the water to catch marine snow particles. Crustaceans known as munnopsid isopods have long hairy arms, many times their body length, which they use to comb the water for marine snow. And perhaps the most splendid snow-catcher of all is the vampire squid. Not a true squid but a distant relative of both squid and octopuses, this blood-red beast is in fact quite gentle and small, no bigger than a rugby ball. Rather than chasing after prey, it unfurls a long filament and waits while snow trickles down onto it. Now and then, the vampire squid reels in its snow-catching device and packs snowballs in its arms before passing them to its mouth and swallowing, with not a drop of blood sucked.

01 – Marine mucilage (also known as sea snot) formed of super-sticky slicks of plankton, linked to warming seas and nutrient pollution

02 – Marine snow falling through the ocean depths

02

11,000 years old
Monorhaphis chuni (glass sponge)

9000 BCE
Giant short-faced bears
and giant ground sloths
go extinct. Mesolithic
people occupied
Britain as the climate
warmed after the end
of the last ice age.

2500 BCE
Stonehenge erected

2181 BCE
End of the Old
Kingdom of Egypt

476 CE
Collapse of the
Roman Empire

650 BCE
Japan founded by
Emperor Jimmu

2,700 years old
Gold corals

15th century
Peak of Aztec and
Inca empires in
the Americas

400 years old
Glass sponges

400 years old
Greenland sharks

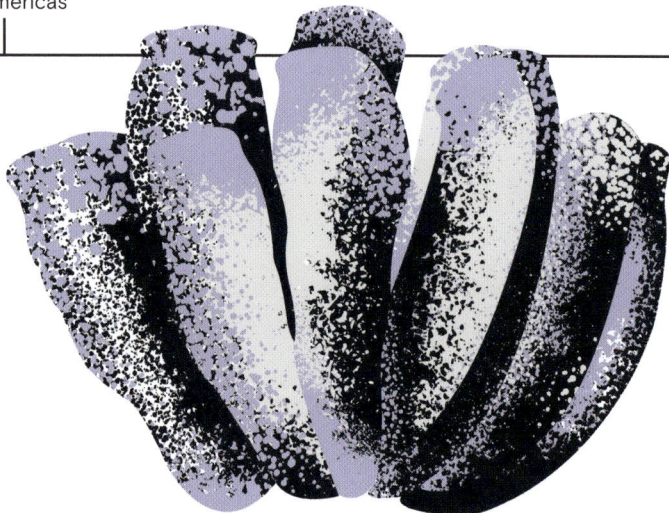

Ancient Life

It is a startling fact that many species have been living in the deep sea for an incredibly long time. This is not simply a case of species being able to trace their direct ancestry far back, as such animals as coelacanths can, but more unusually that individual animals alive today are hundreds or even thousands of years old. They were born all that time ago, perhaps landing on a rock and growing into a sponge or a coral colony, and are still there now.

Why this is so and what quirks of biology allow such immense longevity are puzzles deep-sea scientists are still working on. It may have something to do with the extreme conditions, lack of food and meagre opportunities to find a mate. By hanging around for longer, organisms have a greater chance of successfully dealing with those things, as they play the (very) long game.

Not only are record-breaking deep-sea animals giving us clues to the secrets of eternal youth, but their bodies and skeletons hold important climate records from throughout their immense lives. Scientists can measure tiny traces of certain chemicals laid down by these animals, for instance in the skeletons of deep-sea corals, which can tell us what the temperature, pH and nutrient levels were in the ocean long ago.

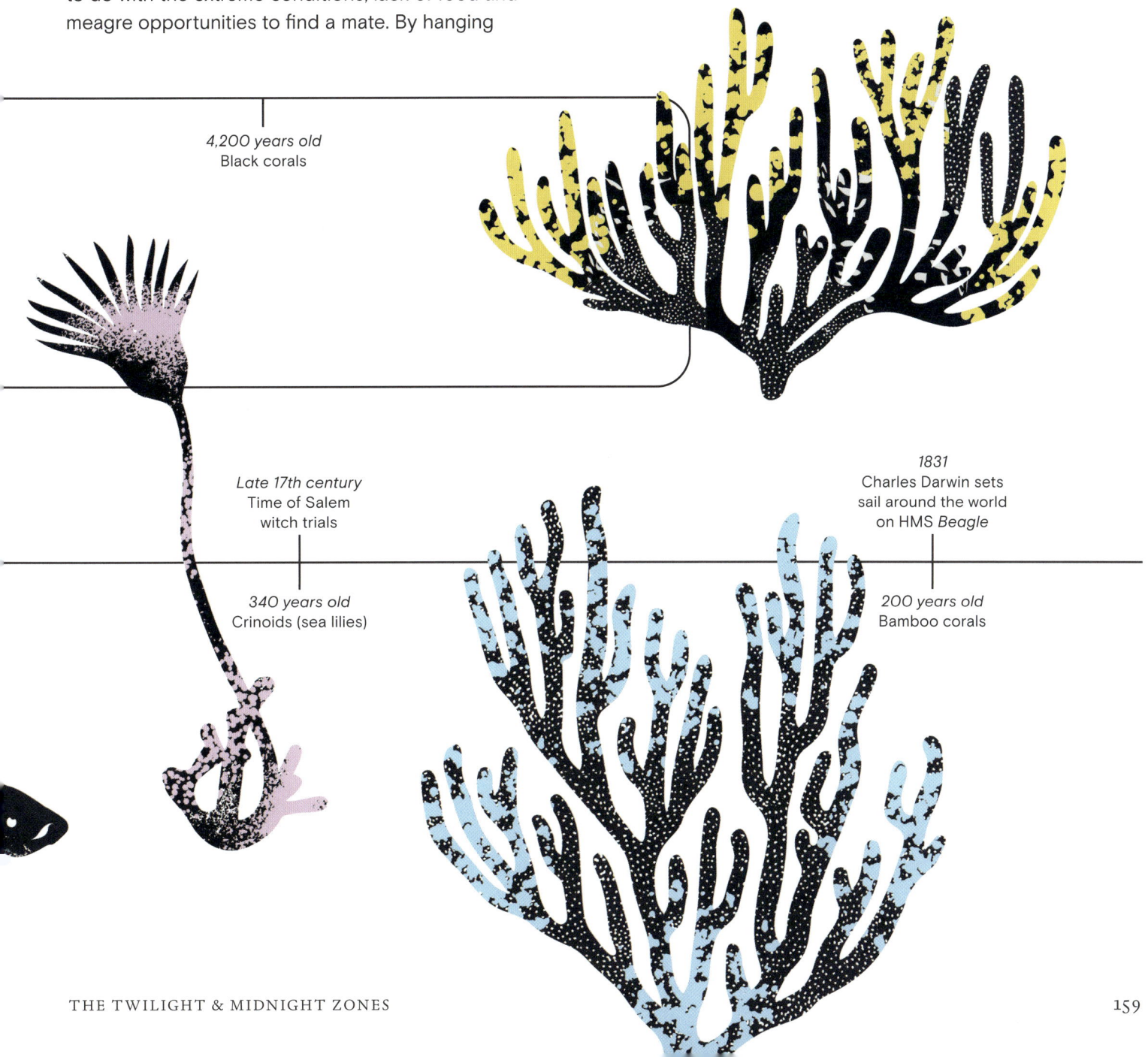

4,200 years old
Black corals

Late 17th century
Time of Salem
witch trials

340 years old
Crinoids (sea lilies)

1831
Charles Darwin sets
sail around the world
on HMS *Beagle*

200 years old
Bamboo corals

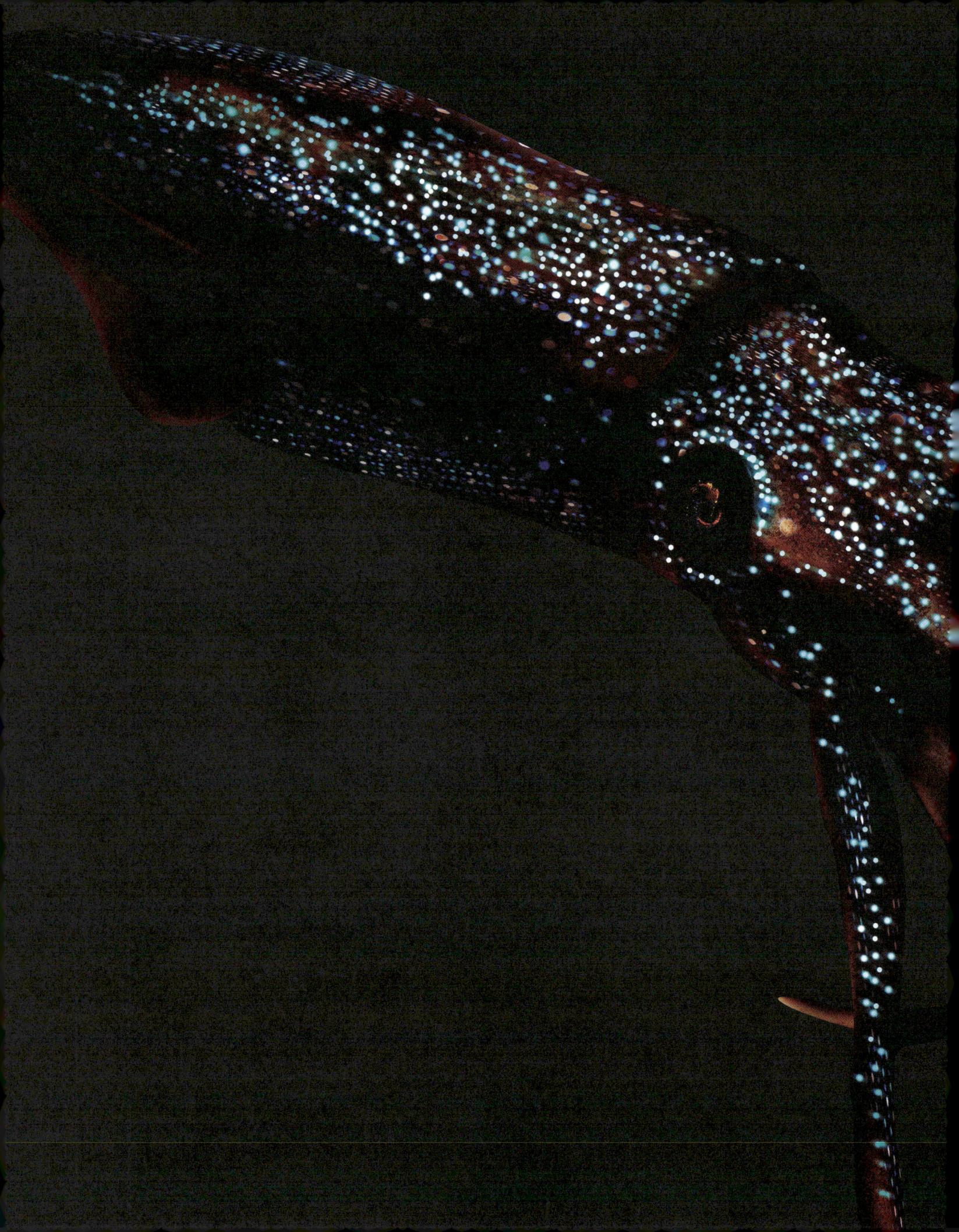

Cloak of Invisibility

In the open waters of the twilight zone, many animals use bioluminescence to conceal themselves when there's nothing to hide behind, a trick known as counter-illumination. Among them are lanternfish, sardine-sized fish that are some of the most abundant vertebrates on the planet and which take part in the nightly migration between the twilight zone and the shallow seas. They illuminate their bellies with dots of blue. Seen from below, these blue belly lights break up the outline of their silhouette, so they no longer look like small, delicious fish swimming by, thereby escaping the attentions of prowling predators. Some species can adjust the intensity of their belly lights as they swim up and down, precisely matching the dim blue light that seeps in from above.

Firefly squid
Watasenia scintillans

Siphonophores

Numerous diaphanous creatures drift through the open waters of the twilight and midnight zones that you could be forgiven for calling jellyfish. Their soft, see-through bodies are made mostly of water. Many have tentacles barbed with poisonous stings that they use to catch and paralyse their prey. And many move with contracting pulses of their rounded bodies, like opening and closing an umbrella. These are not all true jellyfish (members of a group of animals called scyphozoans), but plenty of them are assorted relatives from various nearby branches of the tree of life.

Siphonophores are one such gathering of jellyfish cousins that are especially diverse in the deep sea. Only a single species, the Portuguese man o' war (*Physalia physalis*; see pp. 68–69), is known to float in the shallows. The rest generally live down in the deep, although many participate in the daily vertical migrations and rise up to the surface to feed at night (see p. 180–81).

Especially notable about siphonophores is the way they build their bodies, which blurs the boundaries between individuals and collective colonies. It is often said that siphonophores are made up of hundreds of animals living together, but that is not strictly true. It is more accurate to think of a siphonophore as an animal that rather than growing specialized organs and tissues, such as a stomach and reproductive parts, instead builds itself from tiny clones that each take on a particular role in the body. These clones are called zooids and they all grow from the same fertilized egg. Some zooids are responsible for catching prey with stingers, some digest food, some produce eggs and sperm, and some are the locomotive force that propels the entire colony through the water. None of them could exist on its own and all work together to keep their shared colony alive and to make more siphonophores.

Long before anyone saw a living siphonophore drifting in the deep sea, their shapes and beauty were captured in the drawings of German scientist Ernst Haeckel (see pp. 130–31). He wrote and illustrated the 1888 volume of the HMS *Challenger* report devoted to describing and naming the species of siphonophores the team collected on their epic round-the-world science voyage. Siphonophores also appear in elegant detail in Haeckel's bestselling series *Artforms in Nature* (*Kunstformen der Natur*). Some resemble elaborate bouquets of flowers, some ornate pineapples with ribbon-like tentacles spiralling around them.

01

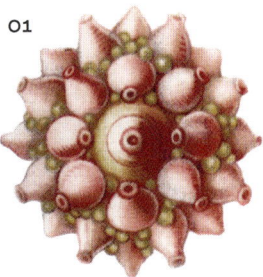

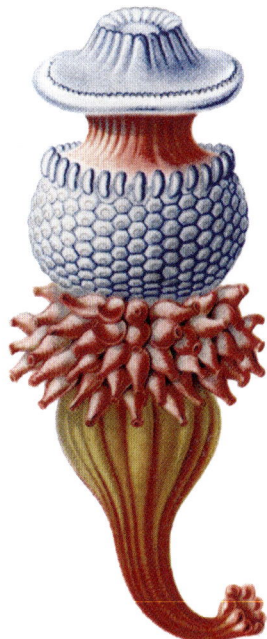

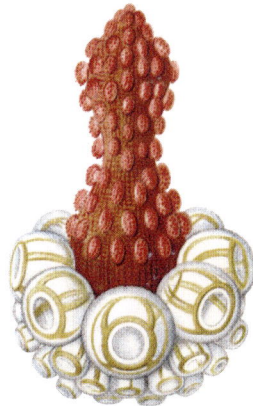

01 and 02 – Illustrations of siphonophores by Ernst Haeckel

Animal Forests

Brittlestar on soft coral

No plants or seaweeds grow in the dark waters of the twilight and midnight zones, but there are animals that live a distinctly plant-like existence fixed to the seabed and together form lush habitats, the deep-sea equivalent of forests. Many underwater forests grow on the flanks and peaks of giant submerged mountains that provide the firm footing these animals need to gain a grip. Corals and sponges are critical in deep-sea forest ecosystems. They grow like shrubs and trees, reaching high into the water column with their lacy branches. Unlike their terrestrial counterparts, these animals are not catching rays of sunlight but grabbing morsels of prey from the water: sponges filter the seawater through their porous bodies, and corals catch plankton with their tiny, petal-like tentacles.

All sorts of other animals make these underwater forests their home. The three-dimensional structures of corals and sponges provide refuge for species to hide away from predators. Brittlestars, long-legged relatives of starfish, wrap themselves around corals. Sea anemones and squat lobsters perch in the branches, catching food from the currents wafting by like a gentle breeze. Octopuses hunt through the forest's undergrowth. And like a forest's canopy, the corals and sponges hoist animals high up above the seabed where there is less chance of getting smothered in falling drifts of marine snow. This is likely why deep-sea catsharks come to these forests to lay their egg cases. Female catsharks entwine their eggs in the branches of corals and leave them dangling like Christmas tree ornaments until they are ready to hatch and the miniature sharks swim away.

Amphipod on *Lophelia* coral

Skate egg case

Common brittlestar

Sea fans with basket star

A Lonely Life

For animals that need a partner to reproduce with, the twilight and midnight zones throw up a big challenge: how to find a mate in the huge three-dimensional darkness. For many species, this part of their life cycle remains a mystery to scientists and for now nobody really knows how they find each other. For some, bioluminescent lights may be involved. For example, velvet belly lanternsharks illuminate particular parts of their body: males have glowing claspers, the reproductive organ that is the shark's equivalent of a penis.

The tactic adopted by anglerfish is also quite clear to see: the males are tiny, but they have big eyes and a heightened sense of smell for finding females, and when they do locate one, they latch on. There are at least 170 species of deep-sea anglerfish and their sexual strategies vary. In some, the males only fix on temporarily and once the female has released her eggs and the male has fertilized them, he disconnects and swims off in the hope of finding another mate. But some anglerfish make this a permanent arrangement: the male's body tissues fuse with his mate's, their blood supplies join together, and he essentially becomes a parasitic bag of sperm. They become bound together for the rest of their lives.

For a female anglerfish, however, the union is not exclusive, and she can collect multiple males clamped onto her body at once. Scientists recently discovered that to avoid rejecting their male hitchhikers and instead accept them as part of their own body, female anglerfish radically alter or entirely switch off genes involved in their immune system. How they survive without their normal immune genes in working order is another unsolved mystery of the deep.

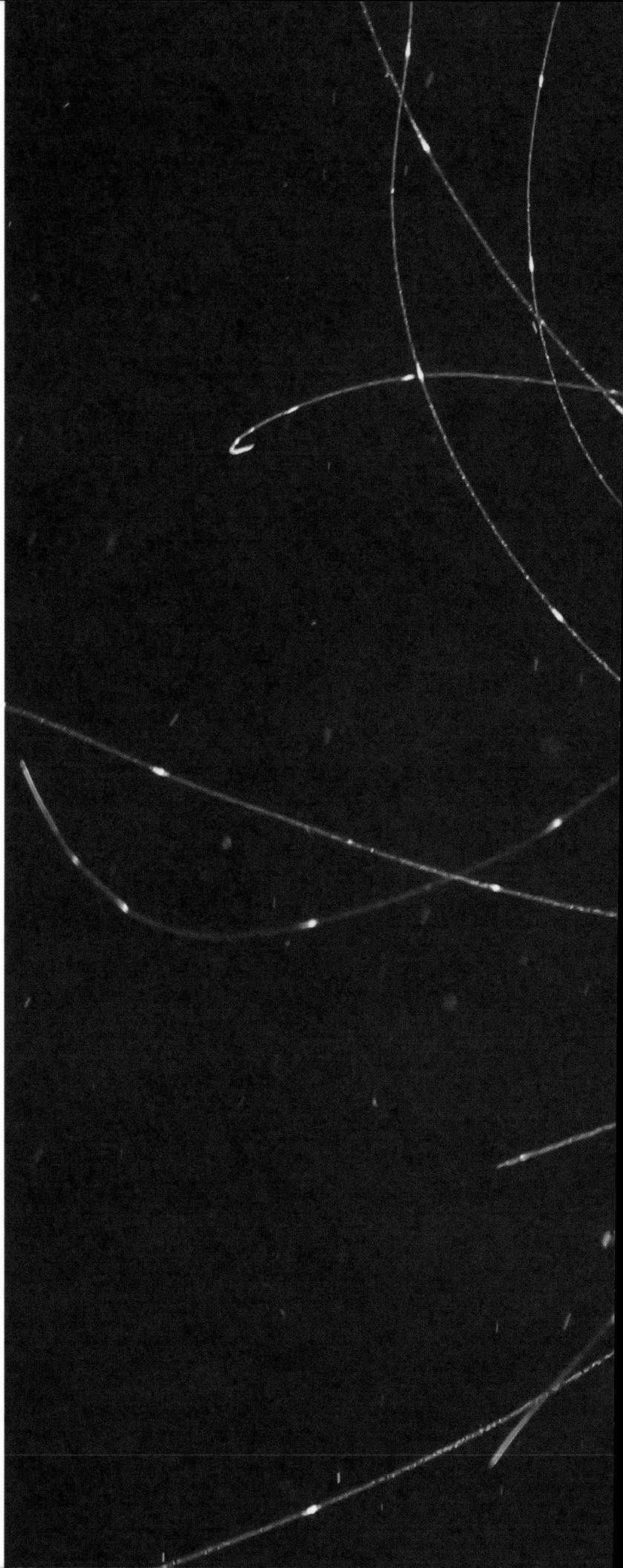

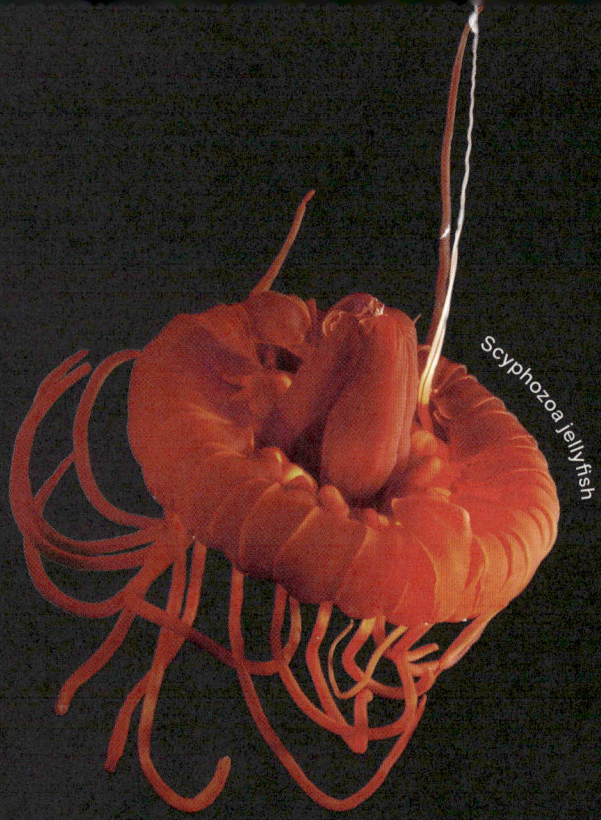

Scyphozoa Jellyfish

Deep-sea mysid shrimp

Deep-sea benthic sea cucumber

Deep-sea red crab

The Colour Red

Red is a common colour among the inhabitants of the twilight and midnight zones. There are blood-red jellies, ruby-red fish, crimson crabs and scarlet shrimp. Among the cephalopods (a subgroup of molluscs that includes octopuses and cuttlefish) there are all sorts of red-tinted varieties, from vampire squid to strawberry squid, the latter appearing extra fruity thanks to the photophores (luminous spots) dotted across their bodies that look like strawberry seeds.

Red is popular in the deep sea because of a lack of ambient red light. Long wavelengths of sunlight at the red end of the spectrum are the first to be absorbed by water. And bioluminescent animals mostly produce blue or green light, sometimes yellow, and almost never red. This means that red pigments appear dull and indistinct in the deep sea because there is no red light available to illuminate them. The colour a pigment appears is determined by which wavelengths it absorbs and which it reflects into the eye of an observer; blue pigments absorb every wavelength except blue, red pigments absorb everything except red, and so on. And so, many animals have found that being red is an excellent way to hide in the deep sea.

Strawberry squid

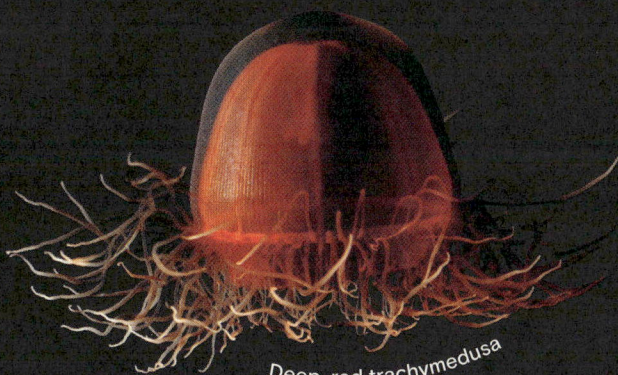

Deep-red trachymedusa

Deep-sea amphipod

Armoured sea cucumber

Living Fossils

Starting in the early nineteenth century, scientists uncovered the fossilized remains of more than 100 species of ancient and distinctly odd-looking fish. They had thick tails shaped into three fleshy lobes, and with no known reports of anyone seeing anything like them alive, people assumed these animals were long gone. In 1839, American ichthyologist Louis Agassiz named them coelacanths (after Latin and Greek words meaning 'hollow' and 'spine', because of the hollow rays in their tail fins). For the next hundred years, fish experts agreed that after evolving around 400 million years ago, the last of the coelacanths had swum through the ocean 70 million years ago, going extinct at roughly the same time as the land-walking dinosaurs.

That view radically changed in 1938 when a fisherman, Hendrik Goosen, hauled in his nets off the coast of South Africa and saw he had caught a strange beast. It was deep steely-blue in colour and around 2 m (6½ ft) long, and had a three-part fleshy tail. Goosen realized this was something special and showed it to a local scientist, Marjorie Courtenay-Latimer. She had the body preserved and began corresponding with J. L. B. Smith, a fish biologist from Rhodes University. Based on the drawings and descriptions she sent, Smith had a hunch as to what this was. And when he paid Courtenay-Latimer and her fish a visit, he knew right away this was indeed a coelacanth that until recently had been alive. He gave it the scientific name *Latimeria*, after Courtenay-Latimer.

The discovery of living coelacanths became a sensation, captivating the public around the world. The finding showed that the ocean holds great secrets and that even large fish can hide out for centuries in the shadowy depths of the twilight zone.

Since finding living coelacanths, scientists have learned a great deal about them. They generally live between 100 and 500 m (330 and 1,650 ft) down and spend the daytime hiding in deep, rocky caves. At night they float about hunting for prey, paddling their large fins in a figure-of-eight pattern, and sometimes adopting an unusual head-down/tail-up pose. They have been found off the eastern coast of the African continent, and off the islands of the Comoros and Madagascar.

In 1997, a second species was spotted by a biologist in a fish market in Indonesia.

The fact that living coelacanths look a lot like their ancestors did hundreds of millions of years ago led to them being labelled 'living fossils'. However, studies of coelacanth DNA have shown that coelacanths have not been stuck in an unchanging evolutionary backwater all this time: in the last 10 million years, they have evolved dozens of new genes. But that did not stop some people latching on to the absurd idea that if you eat a living fossil then you too could live forever (or at least stay looking young for longer). Fortunately for the coelacanth, its meat tastes terrible and has a strong laxative effect, and so a global trade has not taken off. The species are now well protected, although a worrying number of these naturally rare fish are getting caught in nets set to target deep-sea sharks.

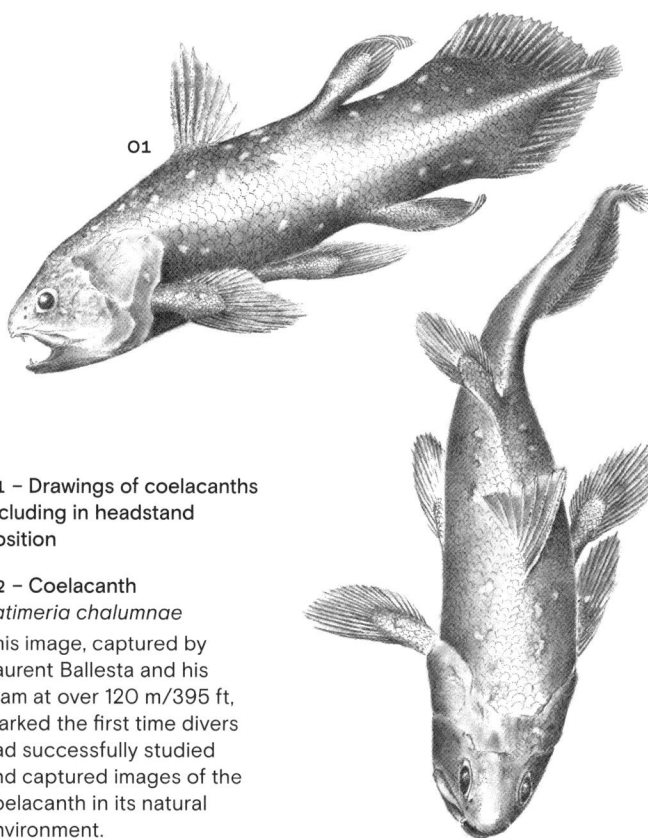

01 – Drawings of coelacanths including in headstand position

02 – Coelacanth
Latimeria chalumnae
This image, captured by Laurent Ballesta and his team at over 120 m/395 ft, marked the first time divers had successfully studied and captured images of the coelacanth in its natural environment.

03 – Fossil coelacanth from the Upper Jurassic in Solnhofen, Germany

Deep-sea amphipod

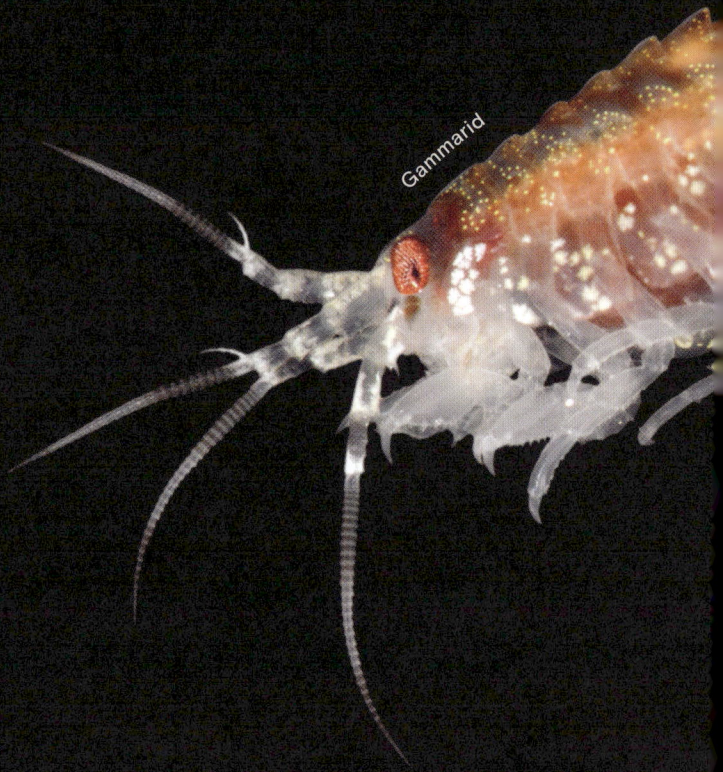

Gammarid

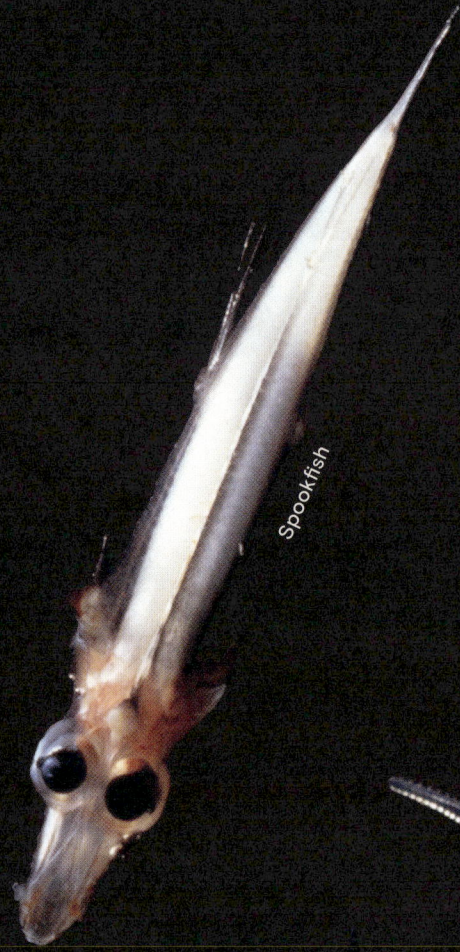

Spookfish

Mesopelagic shrimp

Sci-fi Eyes

Barreleye fish

Amphipod

In sunless caves and down on the deep-sea floor live many animals that have lost their vision. With no light to see by there is no need to devote valuable energy and resources to growing eyes. In stark contrast, the open waters of the twilight zone are home to an array of species with highly sophisticated eyes that catch the last remnants of sunlight trickling down and glimpse the flashes of their glowing neighbours.

Perhaps the most unusual and diverse eyes on the planet are to be found among a group of swimming crustaceans called hyperiid amphipods, distant relatives of the sand hoppers that twitch and wriggle under stones and seaweed at the beach. Hyperiid amphipods are commonly flea-sized, or even smaller, but enlarged under a microscope they look like bizarre science-fiction monsters. Some have eyes that take up their entire head. Some have dozens of individual retinas in each eye to gather as much light as possible. Some have fused their left and right eyes into a single, cone-shaped spike. Scientists are studying these animals to understand why they evolved unique eyes and work out how they perceive their world.

A great many twilight zone species have enormous eyes, all the better for seeing in the dim light. Barreleye fish have long, tubular eyes like a pair of binoculars, which rotate to follow the path of their prey. Strawberry squid, also known as cockeye squid, have mismatched eyes that look in two directions at once: the left eye, huge and yellow, points upwards and scans for the silhouettes of prey passing overhead (the yellow colour filters out down-welling blue light and foils the bioluminescent counter-illumination of many animals that try to blend with the blue ocean surface above; see pp. 160–61); the right eye is smaller and blue, and gazes down to search for bioluminescent animals that sparkle now and then in the dark.

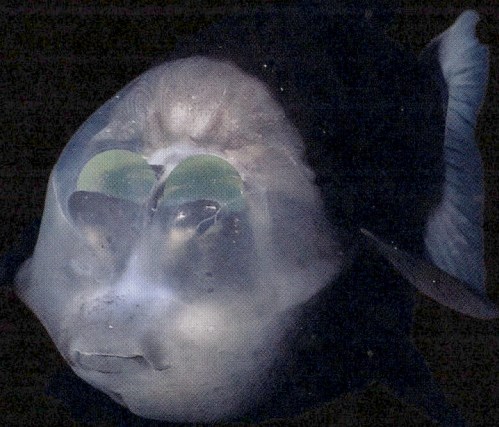

Intricate Sponges

Growing on the seafloor in the twilight and midnight zones are creatures that look like strange fungi or plants, but are in fact simple animals. Sponges are one of the earliest forms of animal life to evolve – fossils of what appear to be sponges have been found dating back more than 800 million years – and today they are especially diverse and abundant in the deep sea. In the late nineteenth century, as scientists began to explore deeper down in the ocean on global expeditions, such as the British team on board HMS *Challenger* (see pp. 260–61) and the German-led team on SS *Valdivia*, they encountered more and more species of sponges that don't live in shallower seas.

Wherever they live, the life of a sponge mostly involves sitting fixed to the bottom of the sea, drawing in seawater through pores in its body (sponges belong to their own animal phylum, Porifera, from the Latin meaning 'pore-bearing') and filtering out tiny specks of food. Some are carnivorous and have sticky spikes and hooks for snagging small prey from the water.

Recent studies using time-lapse cameras on the seabed have revealed that deep-sea sponges can be remarkably active. Some roll around like tumbleweed. And some look like they are sneezing, although it is not surprising that scientists previously missed this phenomenon – it can take weeks to go from an expanding *ahh* to the release of a *choo*. Sneezing sponges are probably dislodging a build-up of unwanted particles from their bodies.

Sponges build their bodies from minute structures, called spicules, made of silica – a basic constituent of glass. Under a microscope the spicules come in all manner of intricate shapes – such as snowflakes, tiny umbrellas and toadstools – that help taxonomists tell one species from another.

In the deep sea, the bodies of sponges come in a huge range of shapes and sizes. They can look like French baguettes, many-pointed stars, candelabras or long socks knitted from fine glass wool. Recently, scientists surveying the deep sea with a remote underwater vehicle came across an enormous white sponge that was larger than a minivan. Tallest of them all is *Monorhaphis chuni*, an astonishingly long-lived sponge, which survives for more than 10,000 years at the end of a long glassy spike as much as 3 m (10 ft) long.

01

01 – Glass sponges

02 – Cloud sponge
Aphrocallistes vastus
Exuma Islands, Bahamas

Lights in the Dark

Even though the sun does not shine in the twilight and midnight zones, it is not completely dark down there. People who have visited these waters in deep-diving submersibles witness fireworks displays created by animals.

Making and controlling light is a tremendous power to possess in near or total darkness. Imagine if you lived in a world without sunlight, but you could light up parts of your body to attract the attention of others, or to use as a torch to see by. Or if you could hurl sparkling particles and slime into the water to distract an attacker, giving you time to disappear into the darkness.

Surveys have shown that roughly three-quarters of species living in the open waters of the deep sea are bioluminescent – they can create their own light. The ability to glow has evolved dozens of times independently in animals dotted all across the animal tree of life, showing just how important this adaptation is for surviving in the deep. There are glowing fish and shrimp, octopuses and squid, worms and jellyfish.

Lights in the deep are produced via chemical reactions. Molecules, either encoded in genes of the light-making animals or within bacteria that live inside their bodies, react with oxygen and release photons of light. Scientists are still learning how animals switch their lights on and off using hormones and nerve signals. Some are prolonged, dim gleams; others are brief, bright sparks.

Blues and greens are the most common colours, with occasional yellow twinkles here and there. These colours can be seen by many animals that have tuned their vision to the blue waters of the twilight zone. One exception is the stoplight loosejaw, a fish that emits its own private wavelength. By producing and seeing red light, it can sneak around in the dark and see things without being seen, as if it were wearing night-vision goggles.

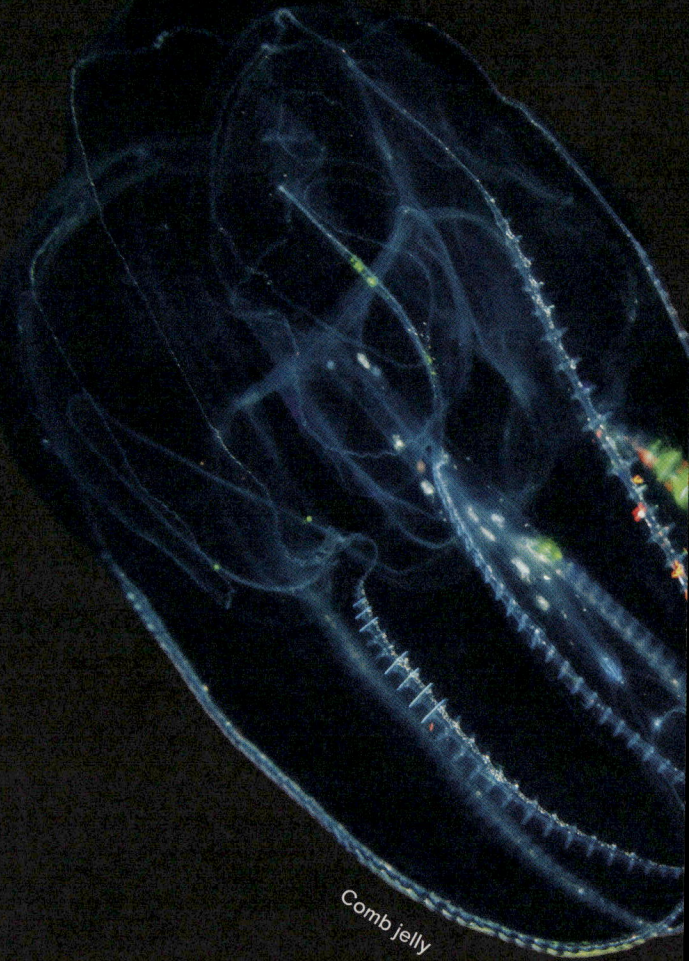

Comb jelly

Comb jelly

Melon comb jelly

Warty comb jellyfish

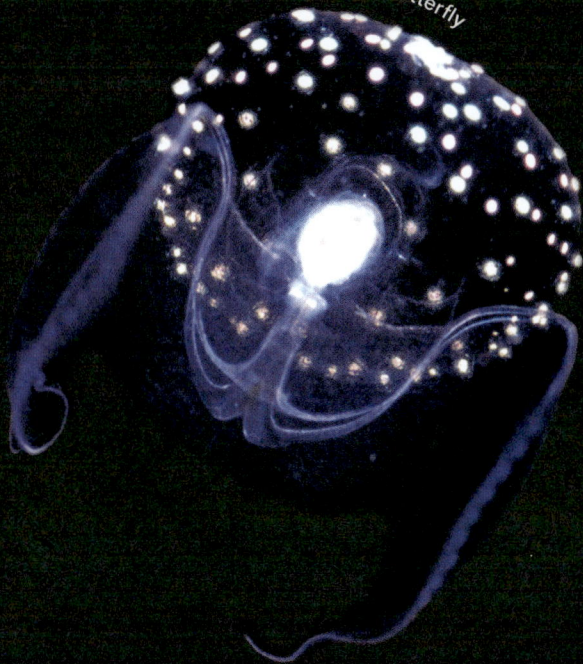

Sea butterfly

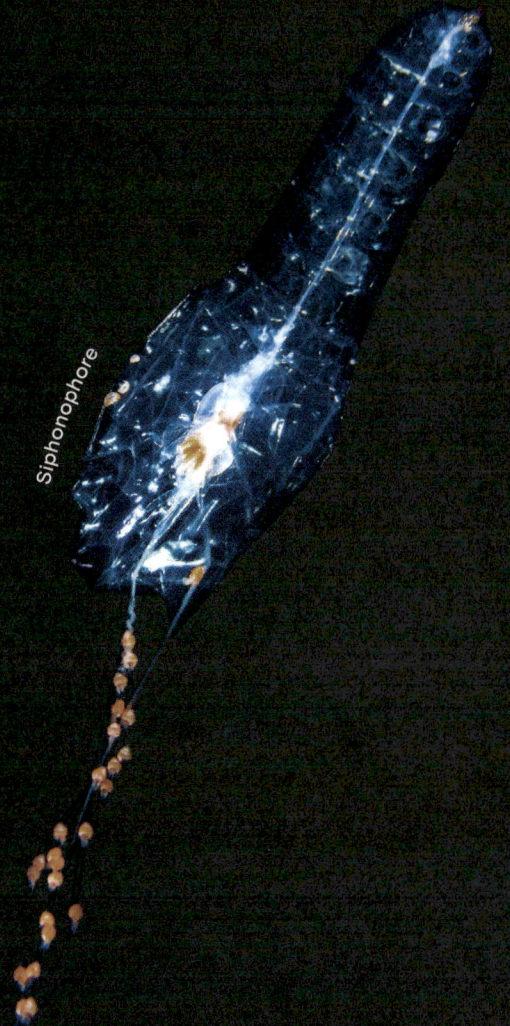

Siphonophore

Helmet Jelly

Rare Sharks

The ocean's most enigmatic sharks spend their lives slinking through the deep sea. Some are known only from the occasional specimens that have been caught in trawl nets and brought to the surface, and have rarely if ever been seen alive in their home waters of twilight and midnight zones. Bramble sharks are aptly named for the prickles that cover their bodies, presumably a defence strategy to make themselves unwelcome prey. Frilled sharks are so named for their ruffled gills on either side of their body. Their sinuous, eel-like body shape harks back to an ancient form of shark that lived millions of years ago during the Carboniferous period.

Many are sharks that glow in the dark (roughly one in ten shark species is bioluminescent), the biggest, at around 2 m (6½ ft), being the kitefin shark. Smaller sharks, such as velvet belly lanternsharks, use their glowing blue bellies to conceal themselves and blend with the dim downwelling light of the twilight zone, but kitefin sharks may use theirs as searchlights to illuminate the seabed as they swim slowly along looking for prey.

At 5–7 m (16–23 ft) long, megamouth sharks are the third-largest fish species in the ocean, after whale sharks and basking sharks, and they are known to roam the twilight zone. As their name suggests, they have huge mouths that take up a large part of their bodies, making them look like giant tadpoles. Like their two larger relatives, they swim with their mouths wide open to sift seawater for plankton. Very little is known about them. People used to think their huge mouths glowed in the dark, perhaps to attract prey, but a 2020 study of skin samples taken from megamouth sharks caught accidentally in Japanese fisheries put that theory to rest, finding no evidence for bioluminescence. They do, however, have a white band of denticles (the tiny toothlike structures that cover shark skin) that shows only when these sharks open their mouths, perhaps reflecting the glowing flashes of plankton. So maybe they do in fact have huge, glowing smiles.

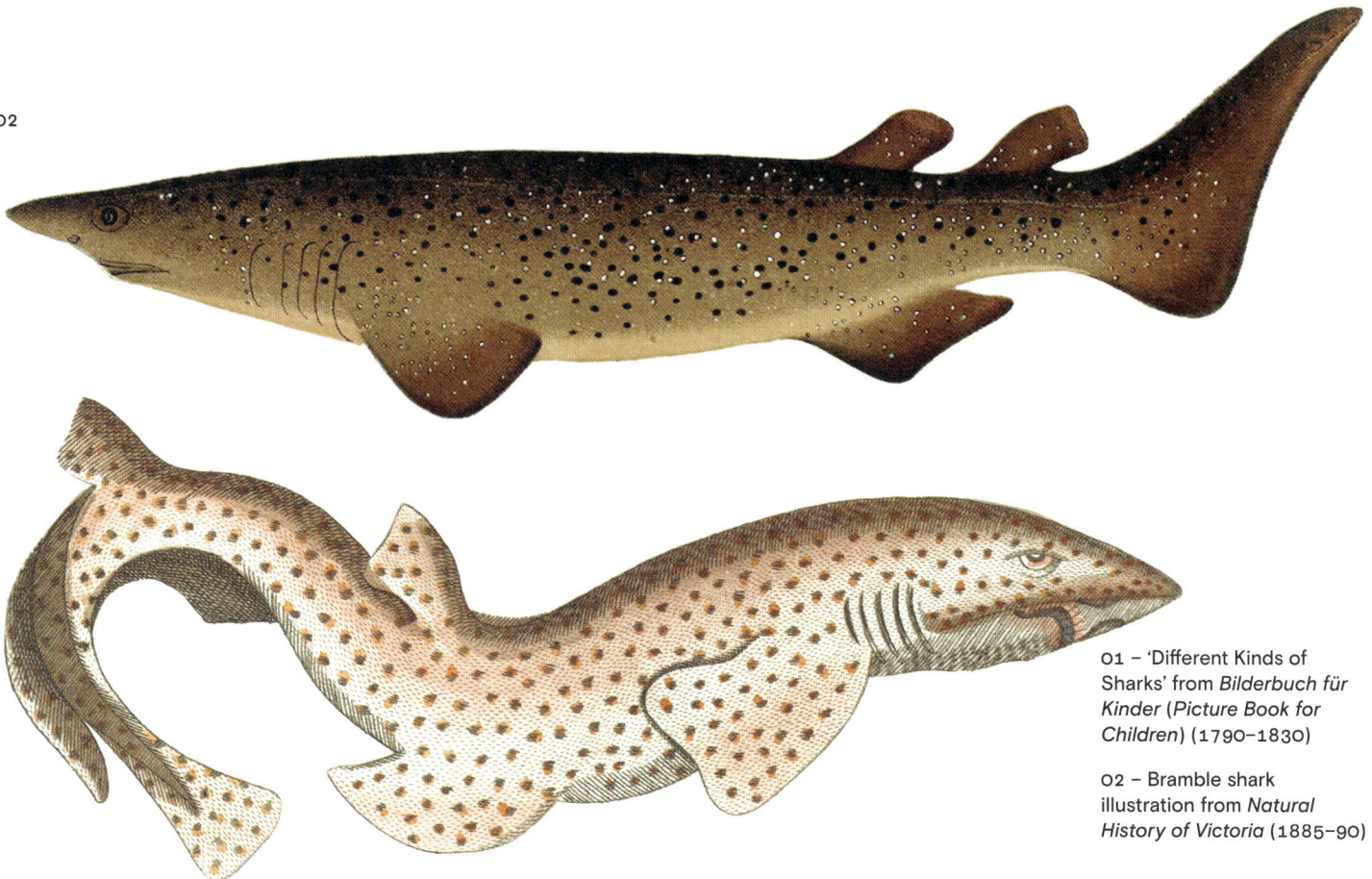

01 – 'Different Kinds of Sharks' from *Bilderbuch für Kinder* (*Picture Book for Children*) (1790–1830)

02 – Bramble shark illustration from *Natural History of Victoria* (1885–90)

Migrating to the Surface

Every night, all across the world, the world's greatest animal migration takes place. Trillions of animals make the journey: squid, fish, shrimp, jellyfish and young larvae of many different kinds. This is not a horizontal migration, as practised by, say, Arctic terns or monarch butterflies, but vertical. As the sun sets, a tremendous wave of animals sets off from the twilight zone and rises thousands of metres towards the surface. They undertake this mass migration to feed on abundant food in the shallows, while avoiding the gaze of daytime predators. The surface seas are much safer at night, when sharp-eyed hunters can't see. Then before dawn, the migrants sink and swim back down to hide once again in the dark depths.

    Scientists have known about this phenomenon since World War II, when US Navy sonar seemed to show the seafloor moving up at night, then sinking back down before sunrise. In fact, the sonar beams were bouncing off the bodies and gas-filled swim bladders of so many animals packed densely together.

    Now, using more refined acoustic sensors and transducers, scientists are learning more about diel vertical migration, as it is known, and the critical role it plays in the ocean's ecosystems and the Earth's climate. Migrating animals consume carbon-rich organic matter in the shallows and shuttle it thousands of metres down, where it will be stored away from the atmosphere for centuries or millennia. Roughly a quarter of all humanity's carbon dioxide emissions are absorbed by life in the ocean, and as much as half of that may be removed by animals that partake in the immense nightly migration connecting the surface ocean to the deep.

01 – Jellyfish *Periphylla periphylla* | 02 – Veined squid *Loligo forbesi* | 03 – Lovely hatchetfish *Argyropelecus aculeatus* | 04 – Krill Euphausiids | 05 – Arrow worm Chaetognath | 06 – Sea butterfly *Clio pyramidata* | 07 – Sharpear enope squid *Ancistrocheirus lesueurii* | 08 – Atolla jellyfish *Atolla wyvillei* | 09 – Isopods Munnopsidae

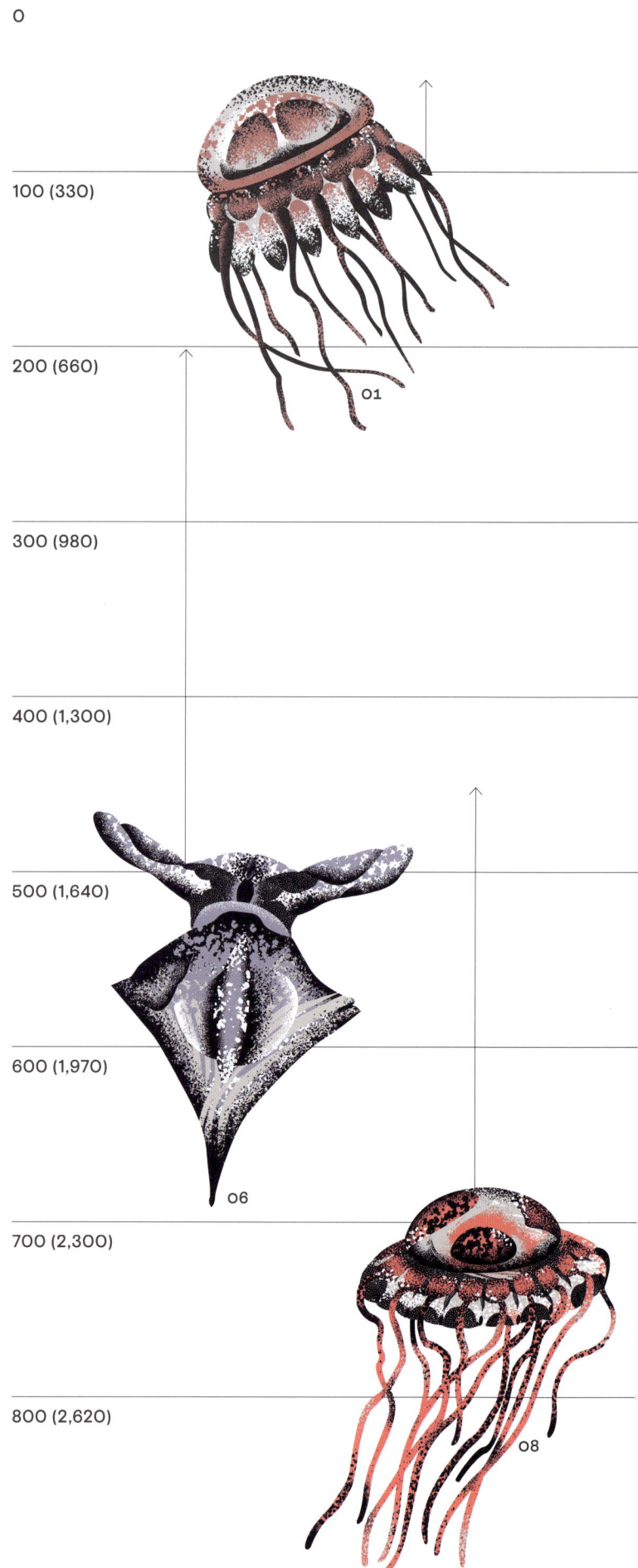

0

100 (330)

200 (660)

01

300 (980)

400 (1,300)

500 (1,640)

600 (1,970)

06

700 (2,300)

800 (2,620)

08

0

100 (330)

200 (660)

300 (980)

400 (1,300)

O2

O3

500 (1,640)

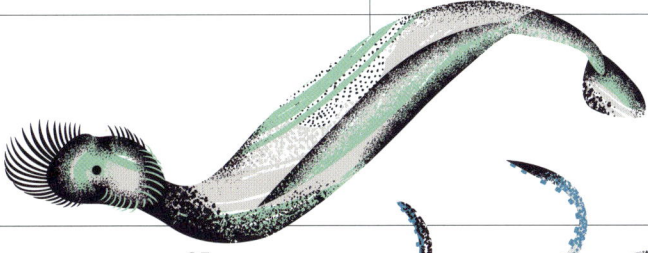

O4

O5

600 (1,970)

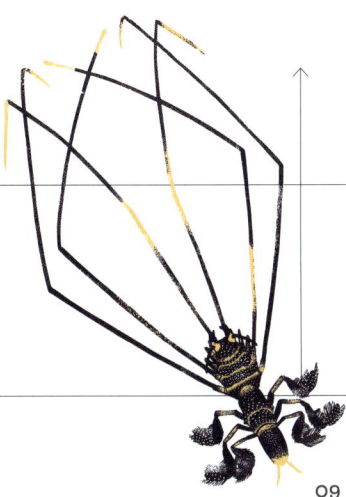

700 (2,300)

800 (2,620)

O9

O7

THE TWILIGHT & MIDNIGHT ZONES

The Abyss

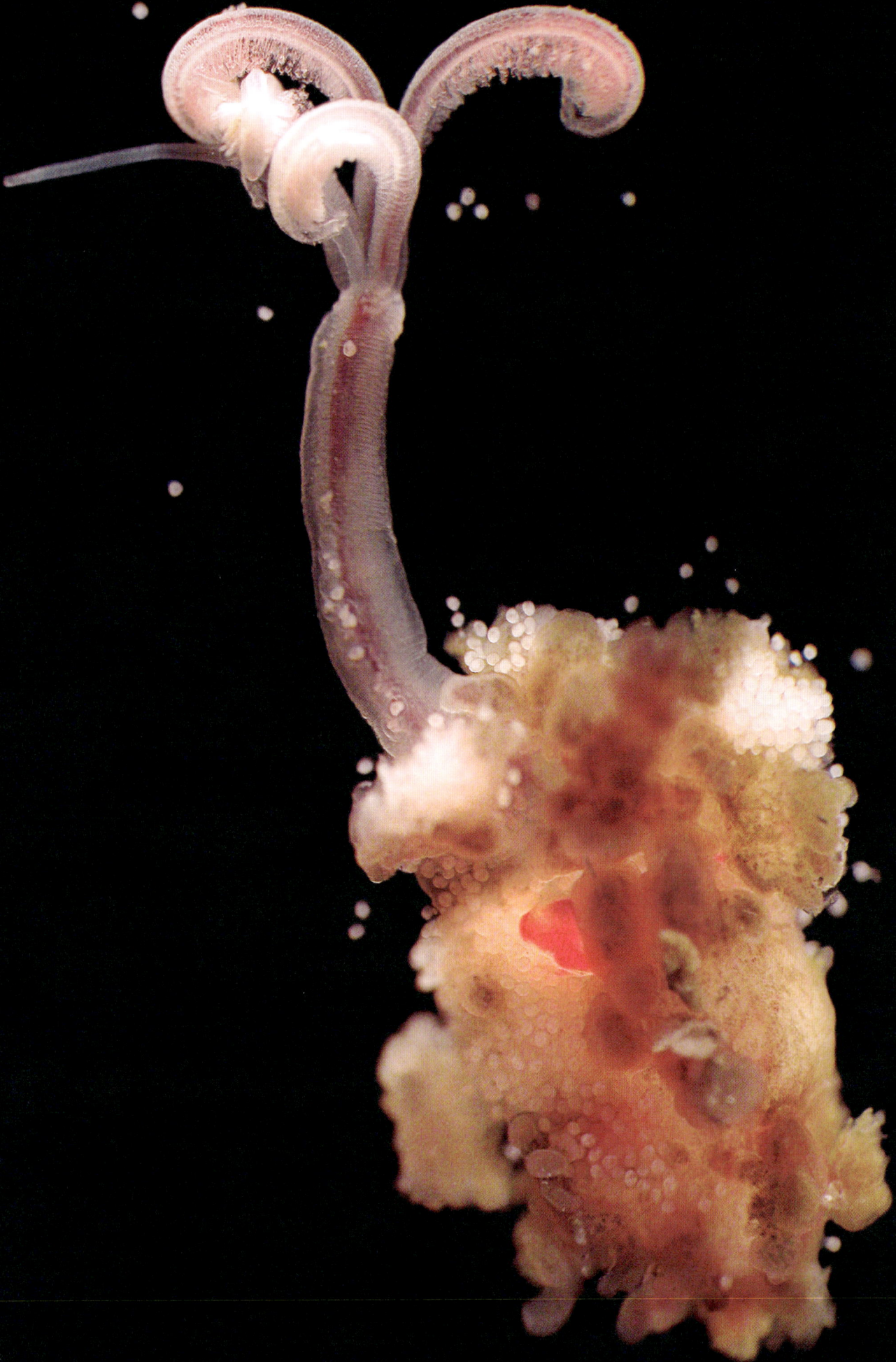

Introduction

The deep sea encompasses the largest and least explored places on our planet. It includes the abyssal plains and remote hadal trenches – ocean regions that dip below the ocean floor to depths of more than 6,000 m (nearly 20,000 ft). This world of darkness, defined by a dearth of food, is called home by animals that have evolved to survive in these extremes. Though seemingly a desolate empire characterized by extreme depth, immense pressure and limited resources, the deep sea contains biodiversity that rivals terrestrial rainforests.

The abyssal plain has been described as 'an empire lacking food'. It is a vast, relatively flat region of the ocean floor found between the continental rise and the oceanic trenches. This rolling seascape of soft sediment and muddy rolling hills extends thousands of metres deep and covers almost 60 per cent of the Earth's surface. The abyssal plain is the largest habitat on the planet. Though once thought to be a biological dead zone, in fact it is teeming with life.

One of the defining features of the deep sea is the absence of sunlight. The sun is the primary source of energy for most ecosystems on Earth, but sunlight does not penetrate the ocean's depths and is nonexistent beyond a few hundred metres. The absence of sunlight means that photosynthesis does not occur. Primary production (the conversion of energy into organic compounds, usually by means of photosynthesis) is rare and strange in abyssal ecosystems.

Without primary production, the abyssal food web is built on alternative energy sources. For many organisms, the primary food source in this ecosystem is marine snow – the continuous shower of organic detritus and dead organisms that sink from the sunlit surface waters (see pp. 156–57). This marine snow provides vital nutrients and energy to the deep-sea organisms inhabiting the abyssal plain. Larger and less frequent influxes to the seafloor in the form of dead animals that sink into the abyss can provide an energy bonanza, creating entire ecosystems built around the afterlife of a dead whale, a fallen fish or even the body of a lost alligator.

Life in the deep oceans must contend with a host of environmental challenges, including crushing pressure, near-total darkness and water temperatures that are barely above freezing. But these challenges also present opportunities. Though faced with extreme pressure, temperature and darkness, the animals that thrive in the deep sea can also count on an ecosystem that is extremely stable, buffered against seasonal changes and radical variability in environmental conditions. The deep ocean is predictable. Except when it is not.

Though often thought of as a vast, barren plain stretching between continents, the deep ocean, like its desert analogue, is freckled with underwater oases that defy the cold, nutrient-poor homogeneity that defines the abyssal plain. Cold-water coral reefs emerge on the fringes of seamounts (submarine landforms). Deep trenches plunge tens of thousands of metres into the ocean crust. Mountain ranges that dwarf anything seen on land wrap around entire oceans, forming the largest contiguous geologic structure on the planet. Mud volcanoes erupt from the seafloor, creating one of the most bizarre geologic spectacles in the sea.

Brine pools, where methane hydrocarbons leach up from deep within the Earth, create lakes on the floor of the ocean. Denser than seawater, chemically enriched salty brine sits on the seafloor, forming puddles, ponds and lakes at the bottom of the ocean. Animals grow around and within these cold seeps in abundance, thriving around an ecosystem that looks, to an outside observer, like an oasis on the seafloor. These methane seeps are influenced by the same forces that shape the surface: brine pools even have tides. Thanks to the intense cold and crushing pressure, a structure called a methane hydrate – a block of frozen methane – forms around some of these cold seeps. Ice worms make their homes within these hydrates, forming burrows from frozen methane. When brought to the surface, these methane hydrates burn like a fistful of natural gas.

Life in the deep oceans must contend with a host of environmental challenges, including crushing pressure, near-total darkness and water temperatures that are barely above freezing.

01

02

Even more stunning are the deep-sea hydrothermal vents. Resembling geysers on the seafloor, these are formed where seawater percolates down through the crust of the Earth, contacts hotspots in the planet's mantle and erupts from chimney-like structures as superheated plumes of metal- and mineral-rich effluent. As these vent plumes contact the cold seawater around them, the minerals come out of solution and are deposited on the walls of the vent, forming an ever-growing chimney of metal-rich rock. With their tangled chimney spires, these vent complexes can grow large enough to dwarf cathedrals.

Species that thrive around hydrothermal vents and methane seeps have taken a novel approach to facing the challenge of a food-limited deep ocean: they eat the vent and seep effluent. Such species as tube worms, snails, mussels and shrimp have evolved to live in the absence of sunlight by tapping into the chemical energy released from deep within the ocean's crust, through a process called chemosynthesis. In chemosynthetic ecosystems, animals have developed their own microbial tools to create food, independent of the sun's energy.

03

Species that thrive around hydrothermal vents and methane seeps have taken a novel approach to facing the challenge of a food-limited deep ocean: they eat the vent and seep effluent.

They form close, symbiotic relationships with bacteria and other single-celled microorganisms that are capable of extracting chemical energy from the effluent and turning it into food for their hosts. Some species have become so dependent on this process that they no longer have digestive systems, relying entirely on their microbial partners to survive.

Where worms, snails, shrimp and fish gather around hot vents and cold seeps, scientists have the opportunity to experience the closest thing to an alien world that anyone alive today will likely ever witness, revelling in the marvels and mysteries of the abyss.

04

01 – Hydrothermal vents found on the undersea volcano Kawio Barat
Indonesia

02 – Deep-water gorgonian coral
Paragorgia arborea

03 – Octocoral

04 – Snailfish
Liparis pulchellus
Slingsby Channel, Queen Charlotte Strait, British Columbia, Canada

The World of Underwater Mountains

Rising thousands of metres from the seafloor, a hidden world of underwater mountains lies beneath the ocean's surface. Known as seamounts, these are often volcanic in origin, forming where tectonic plates crash together or pull apart, or bursting from hotspots in the crust of the Earth. Seamounts provide unique habitats for a diverse array of marine life. They serve as oases of biodiversity in the vast expanse of the deep sea, supporting myriad species ranging from corals and sponges to fish and sharks.

The hard substrate provided by a seamount's rocky slope offers attachment points for a variety of sessile organisms (organisms fixed in one place), creating complex ecosystems. Coral reefs and sponge gardens adorn the slopes of these underwater mountains, providing refuge and feeding grounds for numerous fish species, including many that support commercially important fisheries. The abundance of food and shelter on seamounts also attracts larger predators such as sharks and deep-sea fishes, making them vital hubs of biological activity in the ocean's depths.

In the middle of the southern Atlantic lies one of the more interesting seamount complexes in the ocean. The Rio Grande Rise sits upon the fragments of an ancient continental plate that has drifted, over millions of years, away from the continent of South America. This sunken fragment of an ancient continent is a unique formation of seamounts, interrupting the homogeneity of the vast abyssal Atlantic. It hosts numerous habitats, including cold-water coral reefs, sponge fields, nursery grounds and other hard-bottom habitats.

However, despite their ecological significance, seamounts face threats from human activities such as deep-sea mining and bottom trawling, and from climate change. The fragile ecosystems that thrive on these underwater mountains are particularly vulnerable to disturbance, and the destruction of seamount habitats could have far-reaching consequences for marine biodiversity. Conservation efforts aimed at protecting seamounts and their associated ecosystems are therefore crucial for preserving these unique underwater environments and the myriad species that depend on them for survival. As we continue to explore and understand the importance of seamounts, it becomes increasingly clear that safeguarding these underwater treasures is essential for the health of our oceans.

Yellow ruffle sponge

White trumpet sponge

Stalked white-ruffled sponge

Orange hydroid

Cockscomb coral, growing around an anemone

Galaxy siphonophore

Benthic invertebrates

Bubblegum coral with basket stars

A deep-sea coral community with white tube-dwelling sea anemones

The Iron Snail

On an expedition to hydrothermal vent fields in the Indian Ocean Triple Junction in 2001, scientists discovered one of the strangest snails in the ocean. Clustered in small aggregations at the base of a hydrothermal vent, this snail, *Chrysomallon squamiferum*, appeared to be clad in a suit of armour. Rather than a single hard shell of calcium carbonate, the snail's operculum (the little door that closes when the snail pulls itself entirely into its shell) was covered in a series of tough plates. Even stranger, when the snail was brought to the surface and exposed to air, those plates, as well as the snail's heavy shell, began to rust. This was an iron snail, or scaly-foot gastropod.

This specimen is a truly weird snail. While almost all other gastropods build their shells from calcium carbonate extracted from seawater, the scaly-foot snail extracts iron from the metal-rich plume of the hydrothermal vent, forming a matrix of iron sulphides. The scales that plate its foot and lend it a distinctively armoured look are formed from the same ferrous materials. It is the only known animal that forges its skeleton in iron. This matrix of iron, keratin and calcium carbonate creates a shell that not only is uniquely durable but will actually rust in the air.

Like many vent animals, the scaly-foot snail hosts chemosynthetic microbes that live in symbiosis with the snail, providing food for their host. The uniquely high oxygen demands of the iron snail have produced an extraordinarily oversized heart. At 4 per cent of the creature's total volume, the giant heart of the iron snail is proportionally the largest heart relative to body size of any gastropod, and possibly the largest in the animal kingdom.

This big-hearted iron snail also holds the unfortunate record as the first deep-sea hydrothermal vent species to be listed on the IUCN Red List of Threatened Species, critically endangered owing to the threat posed to its habitat by deep-sea mining.

Hagfish: Beautiful Ugly

Hagfish – ancient, eel-like animals that predate the evolution of vertebrates by several hundred million years – are among the most charismatically ugly creatures of the deep sea. These mobile scavengers have no bones, no jaws and no fins. Their cartilaginous skeleton is composed of a pliable skull, a notochord (the evolutionary precursor of our spine and central nervous system) and a few pliable fin rays. Their skin has been described as 'hanging off of the hagfish like a loose sock'. They barely have eyes.

Hagfish feed by latching onto a carcass using their horizontal, barbed, comb-like mouth parts. They rip off chunks of flesh by tying themselves into knots and pulling backwards against their own bodies. They have a specialized rasping tongue that allows them to rip through the tough outer hides of sunken marine life to reach the soft flesh underneath and are often found burrowing into decomposing carcasses. Hagfish can even absorb dissolved organic matter through their skin and gills.

The hagfish's most remarkable feature, however, is the ability to excrete vast, seemingly inexhaustible, amounts of slime from glands in their skin when threatened, disturbed or, sometimes, perhaps, just bored. Hagfish slime serves as a defence mechanism against predators as well as a tool for scavenging meals. When attacked, hagfish can flood the gullet of their assailant with a thick, viscous slime that clogs the gills and smothers the attacker, giving the hagfish a chance to escape.

Even large sharks are instantly deterred when confronted with a face full of surprise slime. When feeding, the slime creates a barrier, preventing other scavengers from piling in to steal the hagfish's meal. Hagfish in captivity need their tanks constantly cleaned in order to keep the slime from building up.

These are the most primitive of all chordates (animals that possess a spine), and it is highly likely that our ancient ancestors looked and acted like a hagfish. Hagfish Day, a celebration of finding beauty in nature's ugliest creatures, is held on the third Wednesday in October.

01 – Pacific hagfish
Eptatretus stoutii
California, USA

02 – North Atlantic hagfish
Myxine glutinosa
19th-century illustration

03 – Gulf hagfish
Eptatretus springeri
Head portrait. Gulf Specimen Marine Laboratory, Florida, USA. Captive; occurs in Gulf of Mexico

01

O2

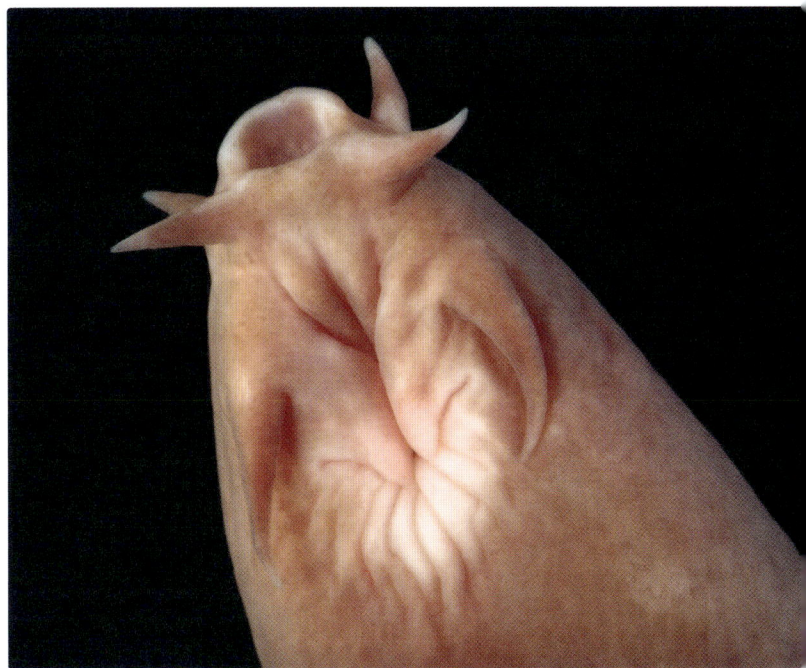

O3

Yeti Crabs

The yeti crab, a subgroup of squat lobsters, is not a yeti. It is also not a crab or a lobster, but belongs to an order of false crabs that includes the more familiar hermit crab, coconut crab and mole crab. Yeti crabs get their name from the silky setae (hairlike filaments) that cover their body and appendages. First discovered in the early 2000s, yeti crabs are found at hydrothermal vents and methane seeps throughout the deep sea.

Thanks to their strange morphology and furry appearance, yeti crabs have delighted deep-sea scientists, who have granted them more whimsical names than those usually assigned to the denizens of the deep. When massive aggregations of a new species of yeti crab were found colonizing hydrothermal vents in the Southern (Antarctic) Ocean, their unique plumage – large tufts of furlike setae emerging from its underside – led to their being dubbed the Hoff crab, in honour of David Hasselhoff's legendary chest hair. Unlike the actor, the Hoff crab probably uses its chest or belly hair to host colonies of sulphur-oxidizing microbes that extract energy from the surrounding vent fluid and convert it into food for the crab.

This trait is not unique to the Hoff crab but is shared among yeti crabs. Another species, lacking a showbiz-inspired common name but bearing the scientific name *Kiwa puravida* in honour of its discovery in Costa Rican waters (*puravida*, or 'pure life', derives from a Costa Rican saying), has extensive setae on its arms. Sulphur-oxidizing microbes have also been found on these setae, which come in contact with chemicals emitted from methane cold seeps. To facilitate this process, the crab waves its claws in the air in what looks like a synchronized dance – the world's deepest crab rave.

Foodfalls in the Ocean Desert

The deep sea is a food-limited desert. Without sunlight to drive photosynthesis, primary production is practically nonexistent outside of a few highly specialized ecosystems. Most deep-sea animals, especially large mobile scavengers such as the giant isopod (a kind of crustacean; see pp. 208–9), are dependent on food from the surface that sinks to the seafloor. This often occurs in the form of organic 'marine snow' – small bits of organic matter that rain down continuously from the surface.

But sometimes, something very big reaches the seafloor.

When a large whale dies and sinks into the deep ocean, the carcass provides a massive influx of food into the deep sea. Large scavengers such as sleeper sharks, cusk eels, hagfish and isopods get to work stripping the carcass of its meat and organs. Depending on how large the carcass is, this process can take years or even decades. Even the skeleton will become fuel for an unusual group of worms called *Osedax* that burrow into the bones and extract the chemical energy from the oil found inside (see p. 228–29). These whalefalls create an oasis of carrion that can persist for decades and support generations of deep-sea communities.

Whales are not the only animals who experience an afterlife on the deep seafloor. Large fish, sharks, sea turtles, dolphins, seals and even alligators and crocodiles occasionally find their way into the deep, where their bodies nourish the animals that call the abyss home. Even woodfalls can play a role, carrying large stockpiles of organic material into the abyss to the benefit of wood-boring worms and snails.

In one notable study in the deep Gulf of Mexico, researchers sank alligator carcasses on the deep abyssal plain. These carcasses were immediately colonized by eels and giant isopods. One carcass, however, mysteriously vanished, dragged away by something large enough to move an entire alligator. The remains were never found and only speculation remains regarding what could have hauled off its hefty body – maybe a sleeper shark?

Long after the meat was devoured, bone-eating worms, previously known only from whale and other marine mammal falls, were discovered colonizing the remaining alligator bones. These worms were a new species within the family of bone-eating worms and their discovery helps scientists better understand the role that ancient marine reptiles, such as the mosasaurs and plesiosaurs that once dominated the sea, played in marine food webs, even after their demise.

Illustration of scavengers
and colonists on a whalefall
by Armando Veve, 2019

Deep Coral Reefs

In a region once believed so ecologically uninteresting that it was viewed as a useful testbed for deep-sea mining equipment, ecologists discovered the world's largest-known cold-water coral reef. Unlike their tropical cousins, cold-water corals live too deep to benefit from sunlight. While tropical corals form symbiotic relationships with photosynthetic microbes (see pp. 108–9), cold-water corals have to feed themselves. Rather than basking in the sun's rays, they filter-feed, capturing food that drifts through the surrounding sea. Like their tropical counterparts, cold-water stony corals grow calcareous skeletons that not only protect the coral polyp inside, but create large, complex structures that form the foundation of an entire ecosystem.

One of the most consistent features of the deep sea is its stability. Most deep-sea ecosystems enjoy an existence of relatively little disturbance, allowing them to persist, unhindered, for millennia. Cold-water coral reefs are no exception. These reefs are long-lived and slow-growing. That stability not only gives cold-water corals the time and space to grow truly massive – some cold-water coral reefs even alter the surface currents – but also makes them incredibly vulnerable to disturbance.

The Blake Plateau is a massive, flat feature with steep walls that lies offshore of the southeast United States. It is one of the biggest singular geologic structures within US waters, and its hard bottom and prominent profile create an ideal space for cold-water corals to settle. Over millions of years, the vast cold-water coral reef, composed of *Desmophyllum pertusum* corals, grew along the western edge of the Blake Plateau, extending from Florida to South Carolina, along 510 km (310 miles) of the plateau and reaching a width of 110 km (70 miles). The 26,000 sq. km (10,000 sq. mile) ecosystem is more extensive than all but two of the USA's national parks and may be among the largest contiguous ecosystems in the continental USA.

Plumose anemone
Metridium senile
White Sea, Karelia, Russia

Pacific Sleeper Sharks

Pacific sleeper sharks are both predators and scavengers, lazily gliding through the deep waters of the northern Pacific and the Arctic as they attempt to expend as little energy as possible while searching for their next meal. Though they are capable of delivering a powerful bite, sleeper sharks prefer to draw their prey into their mouths using suction, allowing them to hunt with little noise or movement. They are excellently stealthy predators, but are not without their own hunters. Orcas prey on sleeper sharks off the Canadian coast.

Very little is known about these enigmatic sharks. The few studies conducted to date have found their bellies filled with giant Pacific octopuses, squid, salmon and even harbour porpoises and sea lions. One shark was caught with a stomach full of giant squid. Their stomachs are unusually large for a shark, allowing them to gorge on food whenever it becomes available. When consuming a large foodfall, they exhibit a characteristic head roll as they tear meat from the carcass.

Though once thought to be confined to the far north, Pacific sleeper sharks have been spotted in the deep western Pacific, near Palau and the Solomon Islands. One sleeper shark was observed living within the warm caldera of an active underwater volcano.

At almost 900 kg (2,000 lb) and 4.4 m (14½ ft) long, Pacific sleeper sharks are the third-largest shark in the ocean, dwarfed only by the Greenland and whale sharks.

01 – Bull shark *Carcharhinus leucas* | 02 – Great hammerhead shark *Sphyrna mokarran* | 03 – Grey reef shark *Carcharhinus amblyrhynchos* | 04 – Great white shark *Carcharodon carcharias* | 05 – Goblin shark *Mitsukurina owstoni* | 06 – Frilled shark *Chlamydoselachus anguineus* | 07 – Sleeper shark Somniosidae | 08 – Greenland shark *Somniosus microcephalus* | 09 – Sixgill shark *Hexanchus griseus*

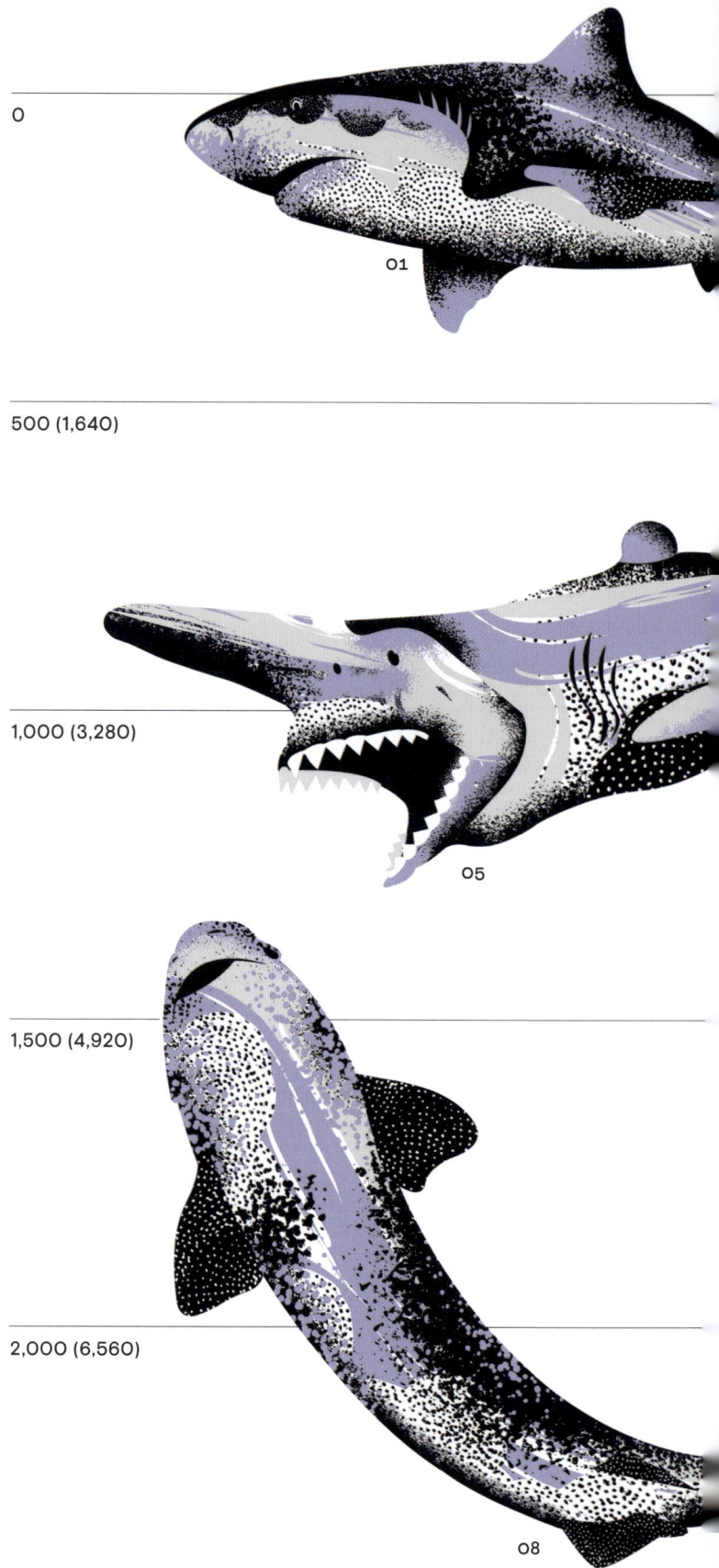

0

500 (1,640)

1,000 (3,280)

1,500 (4,920)

2,000 (6,560)

2,500 (8,200)

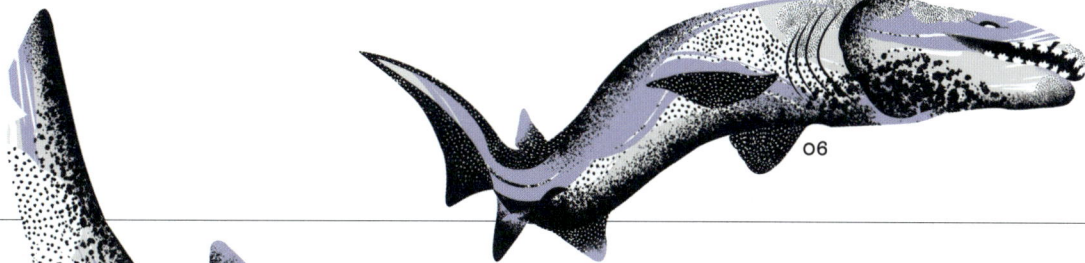

0

500 (1,640)

1,000 (3,280)

1,500 (4,920)

2,000 (6,560)

2,500 (8,200)

Woodlice of the Deep Sea

Imagine a roly-poly, a pill bug, a woodlouse – or any number of common names for the small brown-grey bugs you find underneath rocks and nestled in rotting logs throughout the world. Now imagine a woodlouse as big as a cat. That is the giant deep-sea isopod, whose closest terrestrial relatives are the tiny pill bugs curled up in the undergrowth and decomposing logs of forests around the world.

The giant deep-sea isopod is an enormous woodlouse that inhabits the depths of the Gulf of Mexico. Measuring more than 50 cm (20 in) long and weighing over 1 kg (2 lb), *Bathynomus giganteus* is the largest isopod in the world, crawling across the floor of the Gulf of Mexico and the Caribbean Sea, though some have been spotted scuttling across the abyssal plain as far north as the US state of Georgia, in waters as deep as 2 km (1¼ miles).

Giant deep-sea isopods are adapted to thrive in the dark, cold and inhospitable habitat of the deep abyssal plain. They are scavengers on the seafloor, eating anything that sinks to the bottom of the ocean, including fish, whales and, sometimes, even alligators. Their large size may allow them to capitalize on the boom-and-bust cycle of food on the ocean floor. Foodfalls are rare occurrences, and, thanks to their bulk, giant isopods are able to gorge on any available food. In captivity, giant deep-sea isopods can go for more than two months between meals, making them well suited to the unpredictable availability of food in their remote habitat.

One of the most abundant large scavengers in the deep sea, giant isopods commonly appear as bycatch on deep-sea trawlers, particularly near the Florida coast. Like its tiny cousins, the giant deep-sea isopod can curl up in a ball. It can also swim, gracefully fleeing from large, noisy research robots if disturbed.

Marie Tharp and the Mid-Atlantic Ridge

Marie Tharp (1920–2006) was an American geologist and oceanographic cartographer whose groundbreaking work revolutionized our understanding of the Earth's geology and oceanography. She was the first to map the extent of the Mid-Atlantic Ridge, laying the foundation for the paradigm-shifting discovery of plate tectonics. This fundamentally changed our understanding of planetary geology and answered many mysteries about how species evolved simultaneously on different continents: millions of years previously, the continents had not yet split.

Tharp's early life was marked by a passion for science, despite facing societal barriers against women in academia during the mid-twentieth century. She began her career in the 1940s as a geologist, initially working as a draughtsperson at the Lamont Geological Observatory (now Lamont-Doherty Earth Observatory) at Columbia University, New York. Despite lacking formal academic training in geology, she excelled in her work thanks to her painstaking attention to detail and innate curiosity. She collaborated closely with geologist Bruce Heezen, and together they embarked on a project that would redefine our understanding of the world beneath the ocean's surface.

One of Tharp's most significant achievements came in the 1950s and 1960s, when she meticulously mapped the ocean floor. Through her analysis of seismic data collected from the seabed, Tharp created the first comprehensive map of the Atlantic Ocean's topography. Her maps revealed the presence of a vast underwater mountain range running down the centre of the Atlantic, which provided compelling evidence for the theory of continental drift, a precursor to the modern theory of plate tectonics.

Tharp's pioneering cartographic work not only provided crucial evidence for the theory of plate tectonics, but also paved the way for further exploration and understanding of the Earth's geology and oceanography. Despite being met with initial scepticism and resistance from some in the scientific community, her work ultimately earned widespread recognition and appreciation.

Marie Tharp's legacy extends far beyond her scientific achievements. She broke gender barriers in the male-dominated field of geology, inspiring countless women to pursue careers in science. Her contributions were honoured with numerous awards and accolades, including the Hubbard Medal from the National Geographic Society. Her life and work serve as a testament to the power of perseverance, curiosity and the pursuit of knowledge in advancing our understanding of the natural world.

Tharp's map of the world's oceans, both precise and beautiful, still graces the walls of oceanographers, geographers, geologists and ecologists throughout the world.

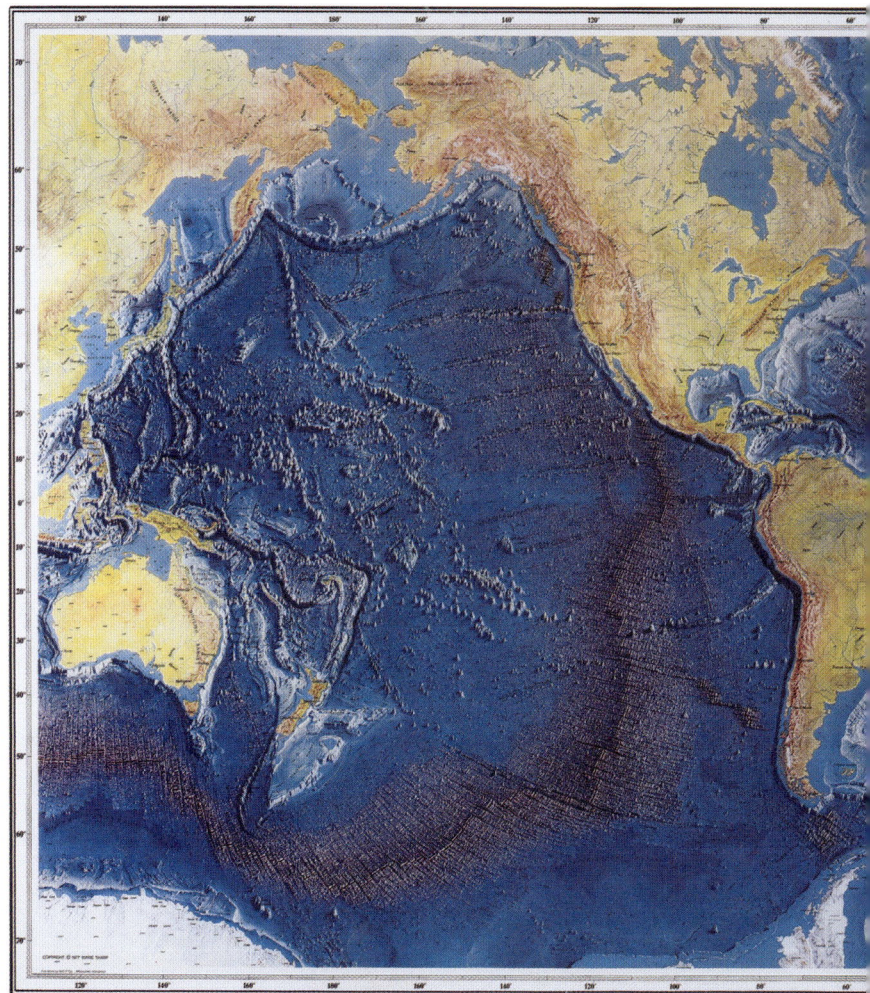

01

O2

01 – World Ocean Floor map
(1977) by Marie Tharp and
Bruce Heezen

02 – Marie Tharp, Al Ballard
and Marty Weiss conversing
aboard the USNS *Kane*

03 – Marie Tharp and Bruce
Heezen at work

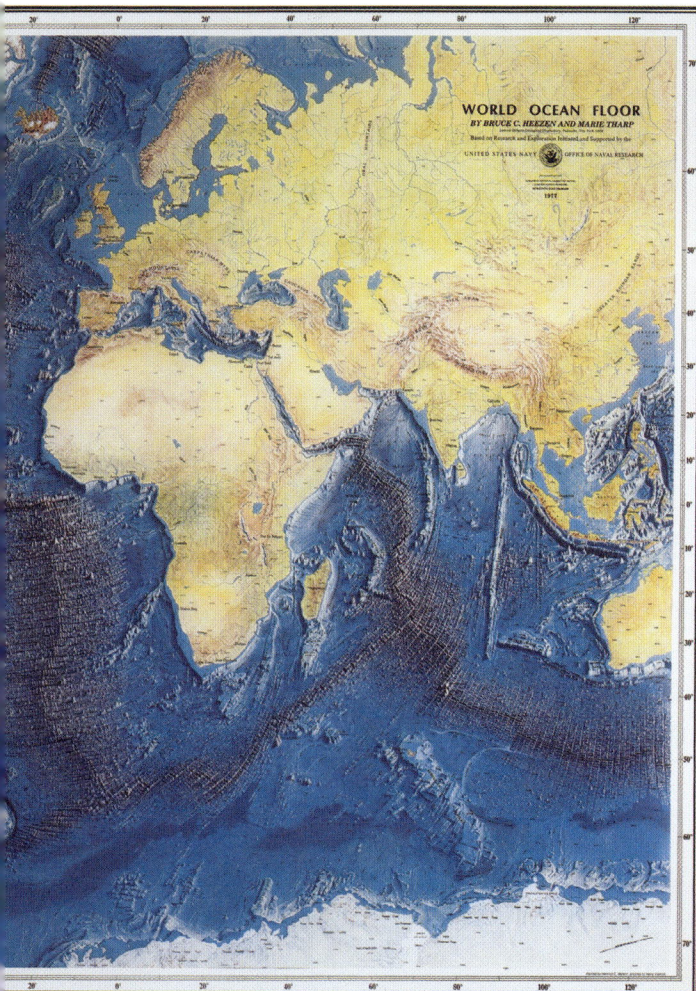

O3

01 – Snailfish
Liparidae
Mid-Atlantic Ridge, North Atlantic Ocean

02 – Dusky snailfish
Liparidae
St Lawrence River, North America

03 – Snailfish
Liparidae
Hiwasa, Tokushima, Japan

01

02

The Deepest Fish

Found throughout the world in the ocean's deepest trenches, the snailfish holds the record for the deepest-living vertebrate on the planet. These remarkable fish have adapted to survive the crushing depths of places as remote as the Mariana Trench – the deepest oceanic trench on Earth. Their gelatinous bodies are naturally buoyant, forgoing the air-filled sac that most fish possess, but which would leave them vulnerable to dramatic changes in pressure. This allows snailfish to navigate the abyssal depths without a swim bladder.

Snailfish species can be found living at depths exceeding 8,000 m (26,000 ft). To survive in this inhospitable environment, snailfish have evolved a unique gelatinous tissue layer that resists the crushing depths and helps maintain their body shape under immense pressure. Most snailfish are blind, having lost the genes that allow photoreceptors to function.

Shaped like an elongated tadpole, snailfish have large, bulbous heads and slender, tapered bodies. They are scaleless, with loose folds of gelatinous skin that hangs off their bodies. This strange layer of skin, which is mostly water, requires very little energy to produce and maintain, allowing the snailfish to grow and thrive in food-limited environments while deterring predators searching for a more energy-dense meal.

Snailfish occupy many ecological niches in the deep sea. As opportunistic feeders, they scavenge on carcasses and small invertebrates, contributing to the recycling of nutrients in the hadal depths. The smallest snailfish live their entire lives within the mantle cavity of a deep-sea scallop, while larger species can be found in kelp forest, in muddy expanses of the abyssal plains, or deep in Norwegian fjords.

Snailfish survival at such extreme depths challenges our understanding of life's limitations and provides insight into how animals, particularly complex creatures, can adapt and survive in some of the most hostile environments on the planet.

Chimaera

Chimaera are sharklike fish that live almost
exclusively in the deep sea, with few observed in
waters shallower than 200 m (650 ft). Also known
as ghost sharks, rat fish or rabbit fish, these ancient
fish are closely related to sharks, skates and rays,
but split from their more numerous cousins around
400 million years ago.

      Like sharks, chimaera possess a
cartilaginous skeleton, but they lack the tough,
toothlike scales that form the skin of sharks and
rays. Rather than long rows of replaceable teeth,
these fish have six large, fused plates that they
use to grind food. Many chimaera also possess the
ability to detect electromagnetic fields through an
electrosensory organ.

      The chimaera's most distinctive feature is its
large, protruding eyes, which give it an otherworldly
appearance. Their large pectoral fins create the
illusion that they are flying through the water like
a bird. Many chimaera also possess a venomous
spine on their dorsal fin.

Rabbit fish, male
Bergen, Hordaland, Norway,
North Atlantic Ocean

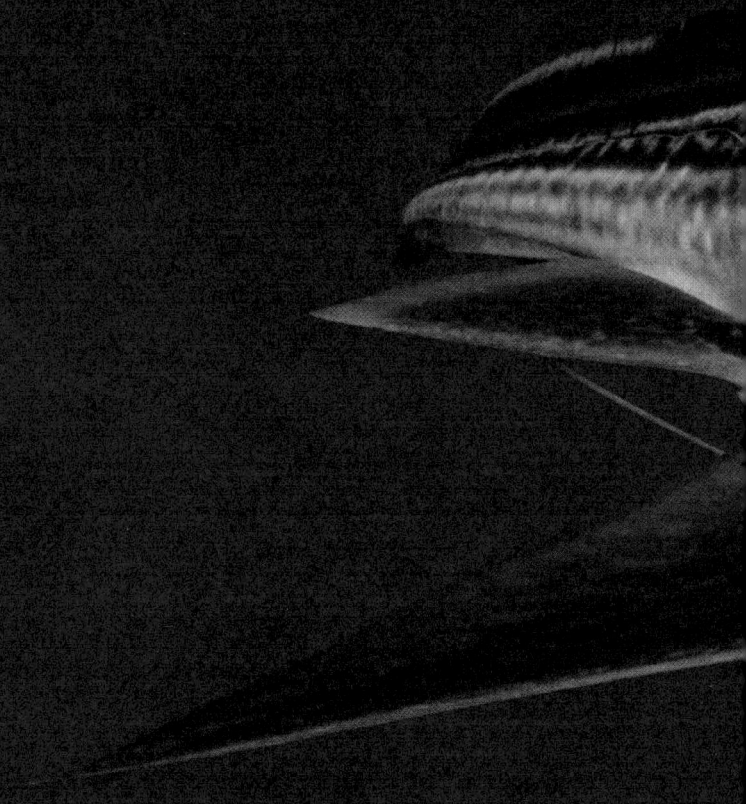

The Cayman Trough

Oceanic trenches – vast chasms sculpted by relentless geologic forces – occur where tectonic plates collide or pull apart, creating fractures in the crust of the Earth. Despite extreme conditions characterized by intense pressure, darkness and chilling temperatures, ocean trenches support unique ecosystems. Exploration of these hadal regions provides insight into the capacity for life in extreme environments.

In the middle of the Caribbean Sea, between Jamaica and the Cayman Islands, lies the Cayman Trough. At 7,600 m (25,000 ft), the Cayman Trough is the deepest oceanic trench in the Caribbean, formed at the boundary between the North American tectonic plate and the Caribbean tectonic plate.

The Cayman Trough, made famous by James Cameron's film *The Abyss*, hosts one of the most unusual and unexpected deep-sea ecosystems. Resting 6,000 m (19,700 ft) beneath the surface is the Beebe Vent Field, the world's deepest-known hydrothermal vent system (see pp. 222–23). Where other vents are populated by massive tube worms, fist-sized snails with shells made of iron (see pp. 196–97), or yeti crabs that dance through the vent effluent (see pp. 200–201), the superheated waters of the Beebe vents teem with *Rimicaris hybisae*, a blind shrimp that lives near the vent in water just cool enough to support it. Although this shrimp cannot see, it has evolved a specialized organ on its shell that can detect the glow of the vent plume. The shrimp needs to do this because it is also a farmer, growing crops of microbes on its gills: and these microbes, via a chemical reaction, produce food for the shrimp.

The ecosystems within the Cayman Trough are remnants of a Caribbean that was once connected to the Pacific Ocean before the isthmus of Panama closed, isolating Caribbean communities from their relatives in the Pacific. Trenches such as the Cayman Trough can provide tremendous insight into how life evolves and adapts to a changing ocean.

Lakes and oceans
Depths and animal/ship/boat lengths are to scale; horizontal distance is not.

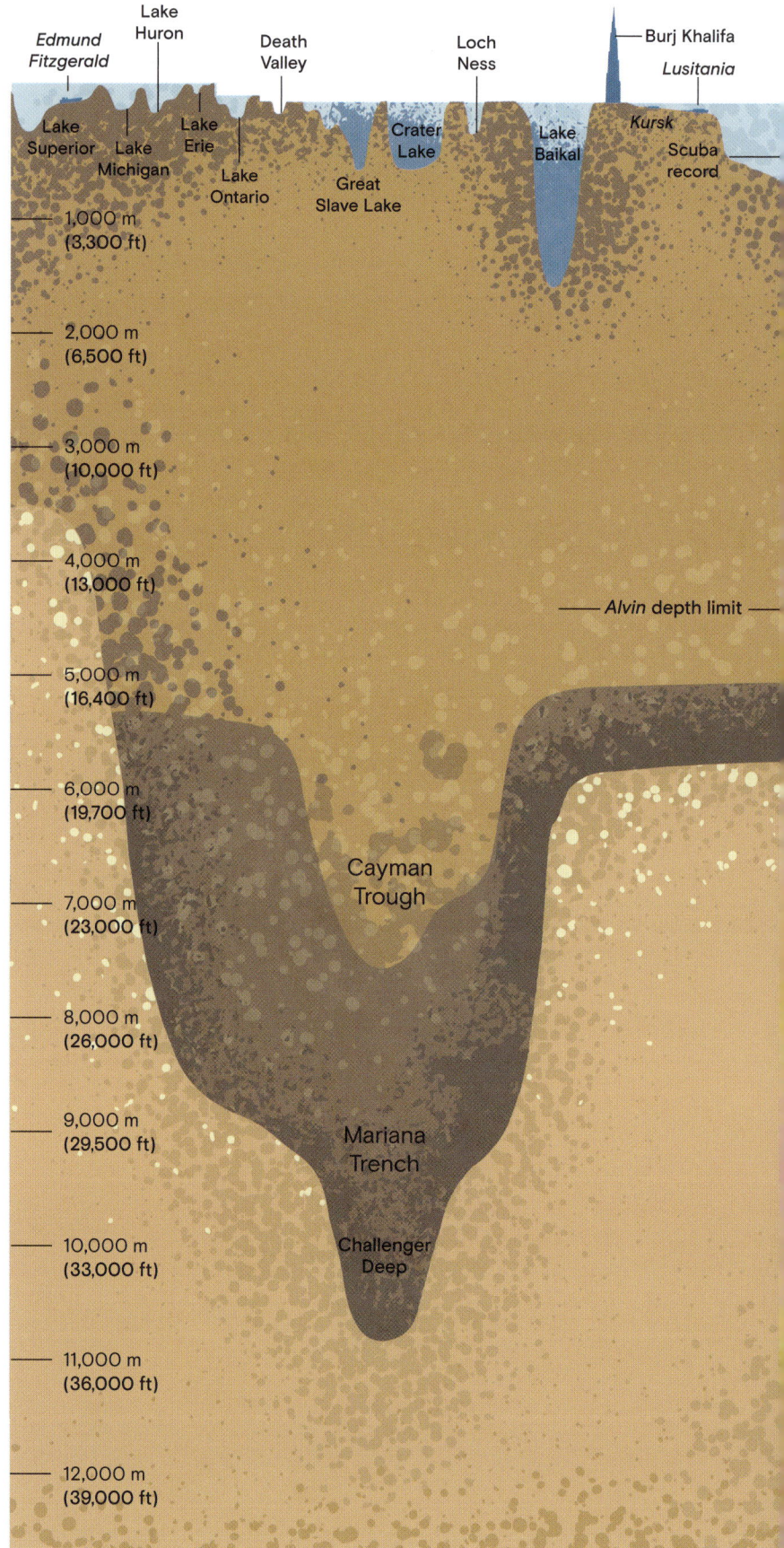

Labels on diagram:
Lake Huron, Edmund Fitzgerald, Death Valley, Loch Ness, Burj Khalifa, Lusitania, Lake Superior, Lake Michigan, Lake Erie, Crater Lake, Kursk, Lake Baikal, Scuba record, Lake Ontario, Great Slave Lake

1,000 m (3,300 ft)
2,000 m (6,500 ft)
3,000 m (10,000 ft)
4,000 m (13,000 ft)
Alvin depth limit
5,000 m (16,400 ft)
6,000 m (19,700 ft)
7,000 m (23,000 ft)
Cayman Trough
8,000 m (26,000 ft)
9,000 m (29,500 ft)
Mariana Trench
10,000 m (33,000 ft)
Challenger Deep
11,000 m (36,000 ft)
12,000 m (39,000 ft)

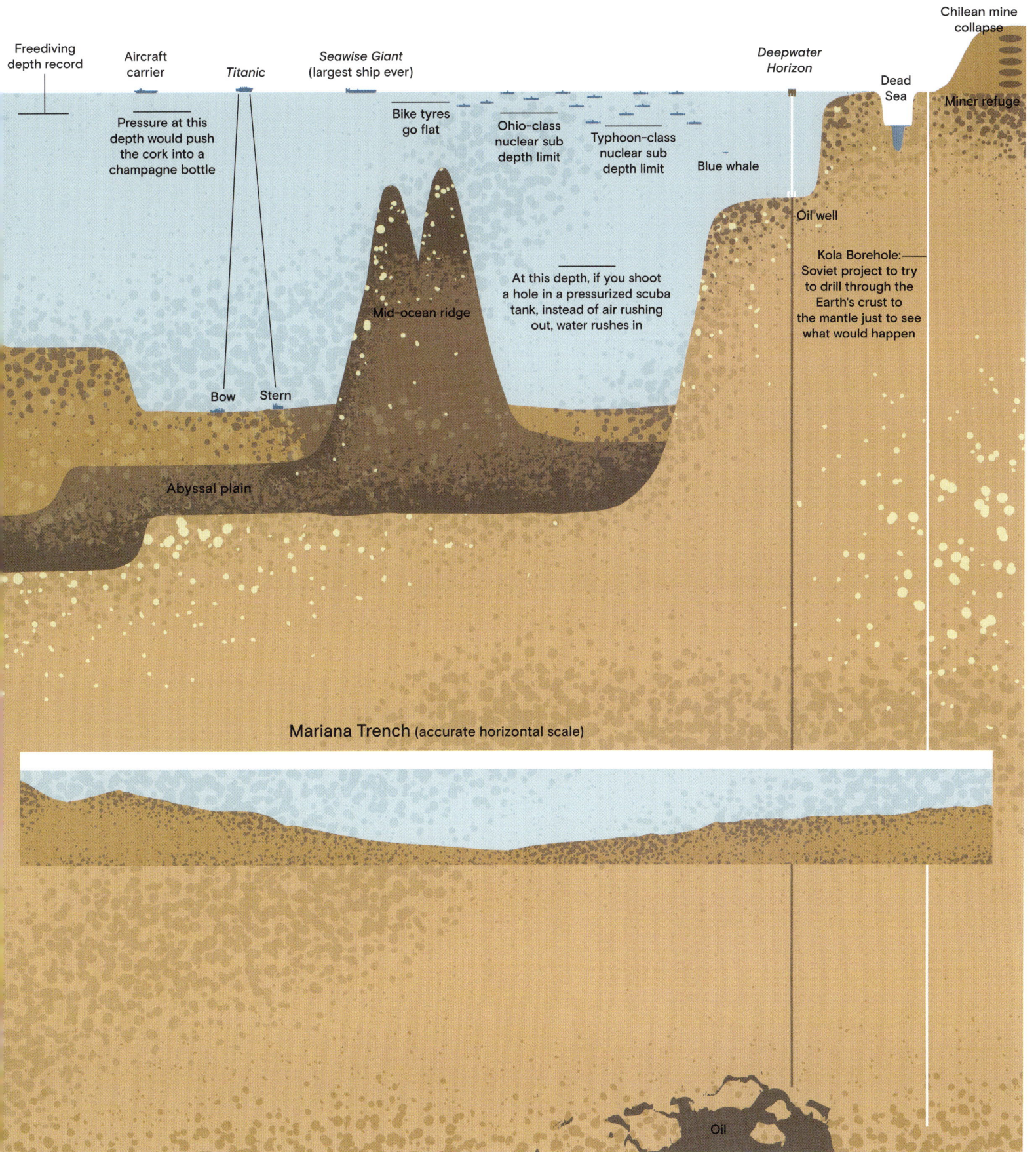

Freediving
depth record

Aircraft
carrier

Titanic

Seawise Giant
(largest ship ever)

*Deepwater
Horizon*

Chilean mine
collapse

Dead
Sea

Miner refuge

Pressure at this
depth would push
the cork into a
champagne bottle

Bike tyres
go flat

Ohio-class
nuclear sub
depth limit

Typhoon-class
nuclear sub
depth limit

Blue whale

Oil well

Mid-ocean ridge

At this depth, if you shoot
a hole in a pressurized scuba
tank, instead of air rushing
out, water rushes in

Kola Borehole:
Soviet project to try
to drill through the
Earth's crust to
the mantle just to see
what would happen

Bow   Stern

Abyssal plain

Mariana Trench (accurate horizontal scale)

Oil

EV *Nautilus*

The Electric Vehicle (EV) *Nautilus* is a modern icon of exploration, discovery and innovation. Named for the submarine from Jules Verne's *Twenty Thousand Leagues Under the Sea*, EV *Nautilus* is the premier research vessel of Ocean Exploration Trust, founded in 2008 by Robert Ballard to 'explore the ocean, seeking out new discoveries in the fields of geology, biology, maritime history, archaeology, and chemistry while pushing the boundaries of education, outreach, and technological innovation'.

EV *Nautilus* began its operational life in 1967 as an East German research vessel named R/V *Alexander von Humboldt*. In 2008, the vessel was transformed into a state-of-the-art exploration vessel. Equipped with cutting-edge technology, including remotely operated vehicles (ROVs), high-definition cameras and sonar mapping systems, *Nautilus* carries equipment capable of exploring depths of up to 6,000 m (nearly 20,000 ft).

Since its maiden voyage, *Nautilus* has embarked on hundreds of expeditions, ranging from exploring ancient shipwrecks to discovering deep-sea ecosystems and geological phenomena. In 2019, the *Nautilus* team, led by Ballard, discovered the wreck of the USS *Indianapolis*, a World War II cruiser sunk by a Japanese submarine in 1945. The discovery provided closure to the families of the sailors lost in the tragedy and contributed to our understanding of maritime archaeology. More recently, *Nautilus* discovered a field of nesting octopuses, which were using the diffuse flow of low-temperature hydrothermal vents to warm their eggs.

EV *Nautilus* is also a platform for education and outreach. Through live-streamed broadcasts and social media engagement, the *Nautilus* team share their discoveries in real time, allowing people around the world to participate in the thrill of exploration. The vessel also hosts students and educators on board, providing them with hands-on experience in oceanography and marine science.

Nautilus pushes the boundaries of ocean exploration. Its ongoing missions seek to unravel the mysteries of the deep sea, from uncovering new species to mapping underwater landscapes. With its dedicated team of scientists, engineers and crew members, EV *Nautilus* remains at the forefront of marine research, inspiring curiosity and encouraging discovery in generations to come.

01

02

03

04

05

06

01 – ROV *Hercules* exploring a crevice in the deep sea

02 – An octopus garden, discovered in Monterey Bay National Marine Sanctuary during a 2018 EV *Nautilus* expedition, and home to more than 1,000 brooding female octopuses (*Muusoctopus robustus*)

03 – A predatory cladorhizid glass sponge, which feeds by capturing and enveloping small prey

04 – ROV *Hercules* sampling a rocky outcropping on the seafloor

05 – A deep-sea Venus flytrap anemone sampled from the Loudoun Seamount, just north of the Midway Atoll in the North Pacific

06 – Gun tub surrounded by glass sponges

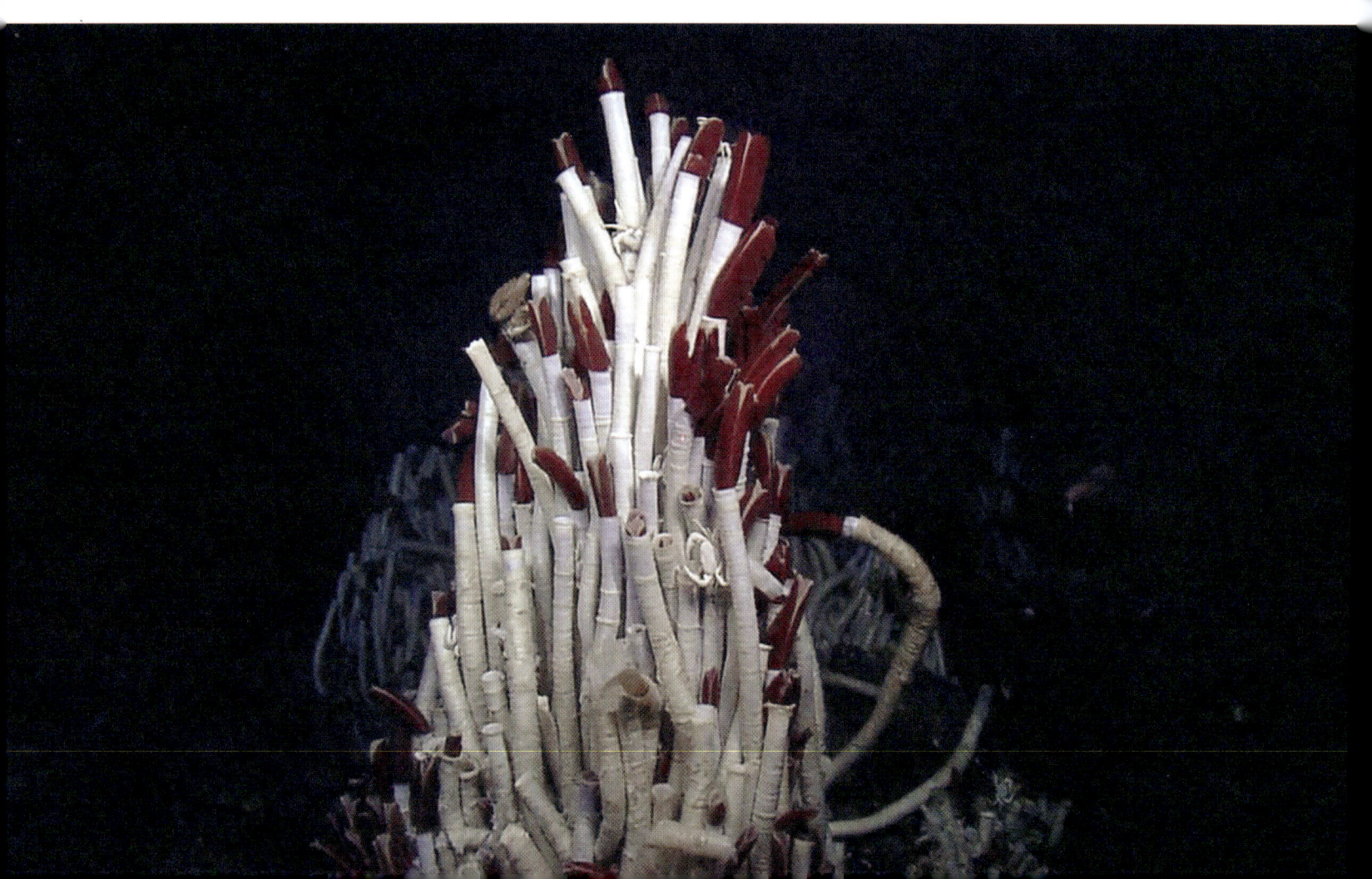

Hydrothermal Vents

In 1977, a research cruise to the East Pacific Rise, a fast-spreading mid-ocean ridge in the northeastern Pacific, made a discovery that fundamentally altered our understanding of what it means to be alive on planet Earth. Deep-sea hydrothermal vents are found where seawater percolating through the ocean's crust becomes superheated and erupts through cracks in the crust of the Earth, creating massive plumes of chemically enriched water. The vents deposit heavy metals and other elements drawn from melting tectonic plates onto the seafloor, forming massive structures that grow into ornate chimneys.

Geologists long suspected that hydrothermal vents existed at continental margins, but they were not expecting life – rich, abundant and unlike anything seen before – to thrive around these vents. So surprising was the first discovery of a hydrothermal vent ecosystem that the team that found it carried no biological preservatives onboard. The first animals ever recovered from a deep-sea hydrothermal vent were fixed in vodka from the chief scientist's personal stash.

It was the giant deep-sea tube worm that captured the world's imagination. These enormous worms build tubes that can reach 3 m (10 ft) in length. From those tubes emerge bright red feathery plumes, which waft through the currents and drift into the chemically enriched waters of the hydrothermal vent. As the colonies of worm tubes grow, they become habitats for crabs, shrimp, squat lobsters, barnacles, octopuses, eel pout fish and hundreds of other species. That first hydrothermal vent sighting revealed so much life that it was named 'Garden of Eden'. Prior to this discovery, we thought all living things depended, ultimately, on energy from the sun to survive. But not these tube worms.

The giant tube worm, *Riftia pachyptila*, is an extraordinary animal. It lives in a symbiotic relationship with microbes that draw chemical energy from the vent plume and convert it into food for the worm, all within a specialized organ inside the tube worm. These worms no longer have a digestive system: instead, they rely entirely on their microbial partners to feed them. The microbes use a process called chemosynthesis to turn chemical energy into organic compounds for the worm to absorb.

The discovery of chemosynthesis sparked a revolution in ecology and evolutionary biology. Once scientists knew what to look for, they began finding it everywhere, including on land: in salt marshes and brine pools, alpine lakes and the deep biosphere – an ecosystem that exists far below the surface, where the heat of the Earth's mantle fuels microbial processes. We find it both in farmers' fields and in the sediment beneath aquaculture pens. A novel, closely related process called radiosynthesis, in which fungi harness ionizing radiation to produce energy, was found in fungi growing within the post-meltdown Chernobyl nuclear power plant.

The tube worm was not the only revolutionary discovery to originate from hydrothermal vents. At hydrothermal vents deep in the ocean and at hot springs on land, scientists isolated a bacterium, *Thermus aquaticus*, capable of producing a protein that allows DNA to replicate at higher temperatures. This discovery led directly to the molecular revolution – the ability to accurately and rapidly sequence and replicate DNA – that forms the foundation of modern medicine.

01 – Brimstone Pit
Hydrothermal venting and release of carbon dioxide bubbles.
Mariana Islands, Northwest Pacific

02 – Giant tube worms
Riftia pachyptila
Deep-sea hydrothermal vent tube worms rising from a vent chimney

Spycraft

Humans have found many ways to use and misuse the deep ocean, but perhaps its most unexpected appropriation is as a place of espionage. The vast expense of deep-ocean exploration means that partnerships between research, exploration and military activities are frequently the norm rather than the exception. Amid the conflict of the Cold War, the deep ocean served as a clandestine battleground between the United States and the Soviet Union.

Two famous covert operations are carved into the history of deep-sea exploration. Project Azorian was a CIA-funded operation to recover a missing Soviet submarine that sank in the middle of the Pacific, in waters 6,000 m (nearly 20,000 ft) deep. Using billionaire Howard Hughes's emerging interest in deep-sea mining as cover, the Central Intelligence Agency constructed the *Hughes Glomar Explorer*, a mining vessel whose hull hid a large central docking well (known as a moon pool), a clawlike capture vehicle and a raising system capable of lifting a submarine into its hold without being observed. Unfortunately, less than half of the submarine was recovered, and no useful information was obtained. The mission itself is considered one of the most expensive intelligence failures in US history. But the *Glomar Explorer* remained in use by US aerospace company Lockheed Martin as an exploratory vessel for deep-sea mining. The operation was declassified in 2010.

More recently, famed deep-sea explorer Robert Ballard revealed that his search for the *Titanic* was actually a military-funded expedition to examine the wreckage of two US submarines, the *Scorpion* and the *Thresher*, which were lost in the Atlantic in the area near where the *Titanic* went down. Hunting for the famous ocean liner served as cover so as not to reveal the location of the two submarines, their nuclear payloads and the intelligence information that they contained. That mission was a success, both submarines were surveyed, and Ballard also succeeded in finding the *Titanic*.

The discovery of the *Titanic*'s wreckage was in fact inconvenient for the US Navy, as they were then obliged to support subsequent expeditions to its site in order to maintain the cover story.

01 – Electric submarine
Claude Goubet (1886)

02 – USS *Scorpion*, a nuclear submarine that sank in May 1968 for reasons still uncertain

03 – Tabula Decimaquarta, *De Motu Animalium*, vol. 1 by Giovanni A. Borelli (1680)
Illustrations of nine marine and submarine navigational and breathing devices

01

O2

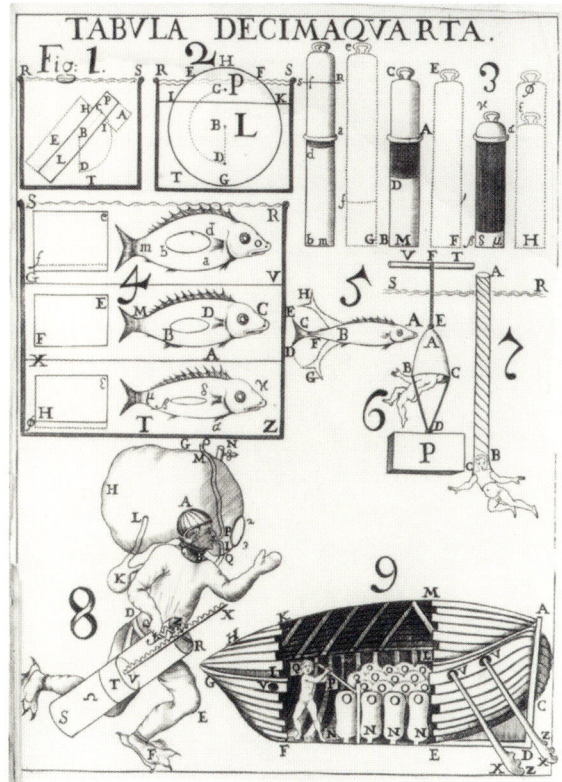

O3

Polymetallic Nodules

Scattered across the abyssal plain, covering thousands of square kilometres of seafloor, are vast fields of polymetallic nodules. These small stones are among the most unusual geologic formations in the ocean. Each nodule grows around a 'nucleating agent', often the hard shell of a diatom (a kind of algae) or fragments of detritus that have drifted down from the surface, but sometimes the tooth of a shark or the bone of a whale, which can be found preserved within the nodule. The geologic skin of the nodule is formed from heavy metals suspended in the seawater, which are slowly deposited over millions of years. A potato-sized rock on the seafloor might have taken 10 million years to form.

Polymetallic nodules grow in fields that can dwarf other ecosystems. The Clarion Clipperton Zone in the North Pacific is a nodule field almost as large as Australia. The nodule fields, which look like acres of scattered cobblestones, may in fact be the largest contiguous ecosystem on the planet. Though they are energy-poor ecosystems where biomass is limited by available food, the biodiversity of these nodule fields is tremendous, with some estimates suggesting that the biodiversity of polymetallic nodule fields on the abyssal plain could rival that of tropical rainforests.

Nodule fields support communities that rely on the hard surfaces provided by the nodules to survive. Sponges, corals, worms and barnacles have all been observed growing on these nodules. The nodules are valuable to humans as well. A polymetallic nodule is rich in manganese, cobalt, nickel, copper and elements rarely found on Earth, many of which are key components in the batteries of electric vehicles. This vast and unaccessed deposit of critical metals has led some to begin exploring the potential of mining the deep sea.

Curiously, while nodules are found on the seafloor, they are not found buried within the sediment. Geologists still do not understand the precise mechanism by which these nodules form, but some theories suggest that living organisms may play a part in the process. Microbes acting on minerals may break down nodules that sink into the sediment. Alternatively, the movement of currents over the abyssal plain or the burrowing action of the creatures of the seafloor may maintain the nodules' position on the seafloor over millennia.

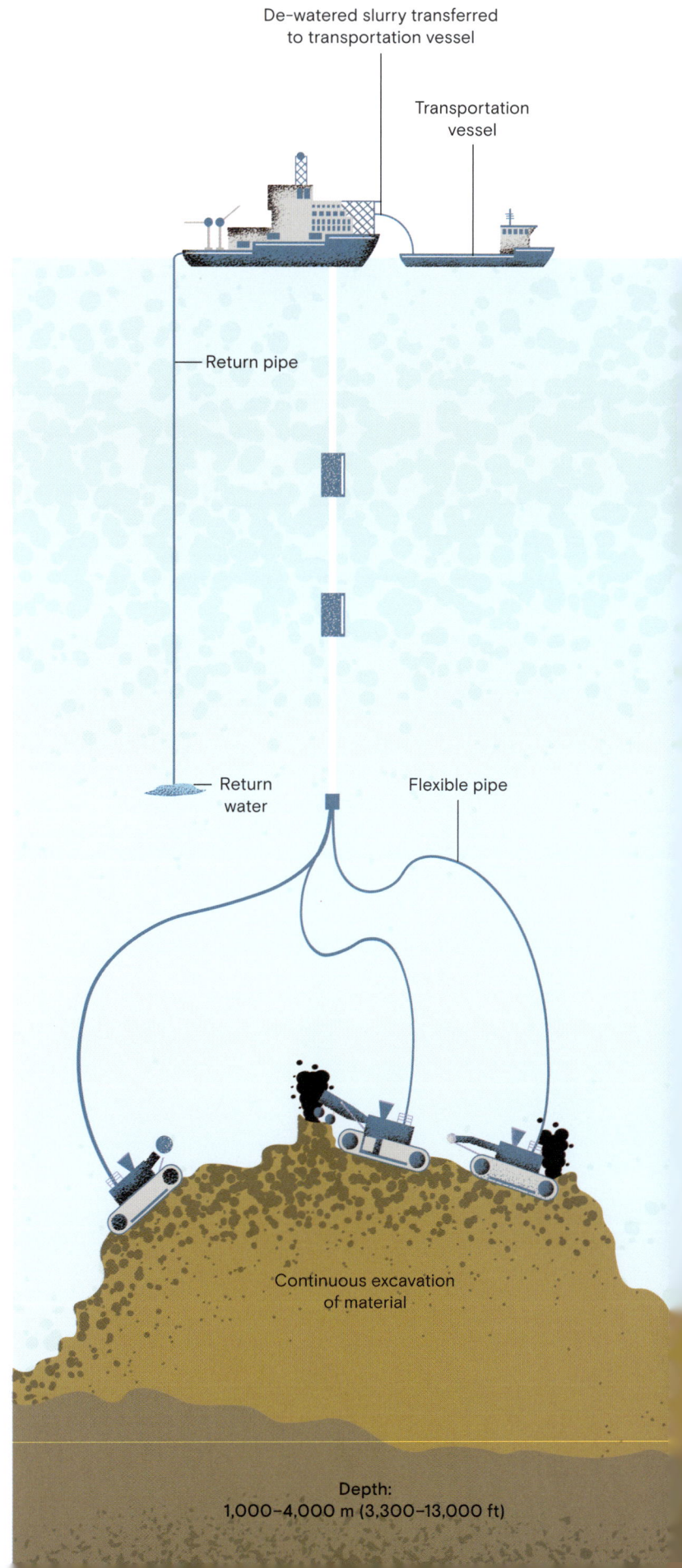

Seafloor massive sulphides on active and inactive hydrothermal vents

De-watered slurry transferred to transportation vessel

Transportation vessel

Return pipe

Return water

Flexible pipe

Continuous excavation of material

Depth:
1,000–4,000 m (3,300–13,000 ft)

Polymetallic nodules
on abyssal plains

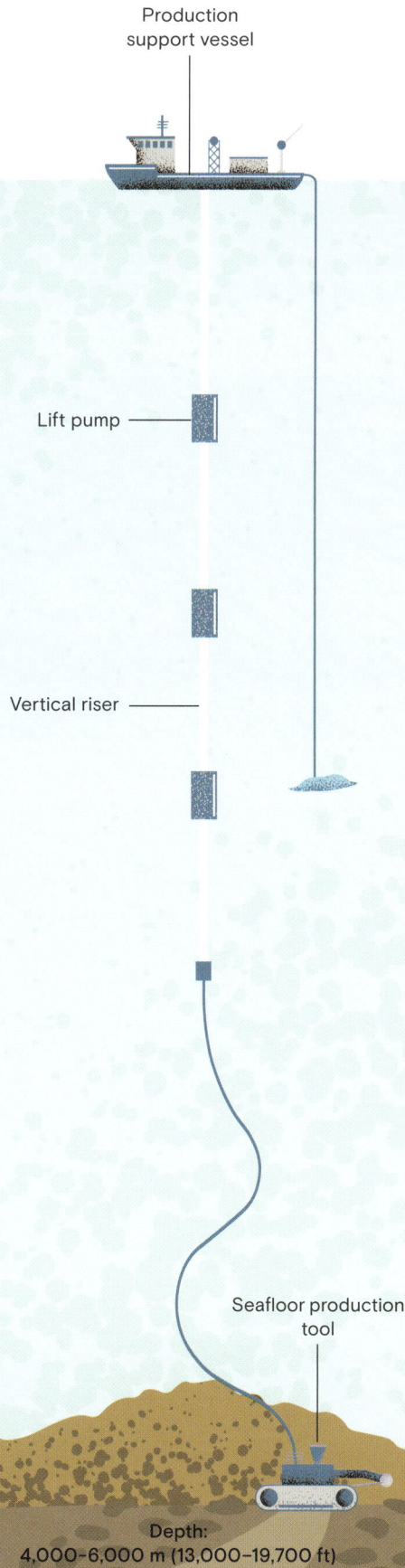

Cobalt-rich crusts on slopes
and summits of seamounts

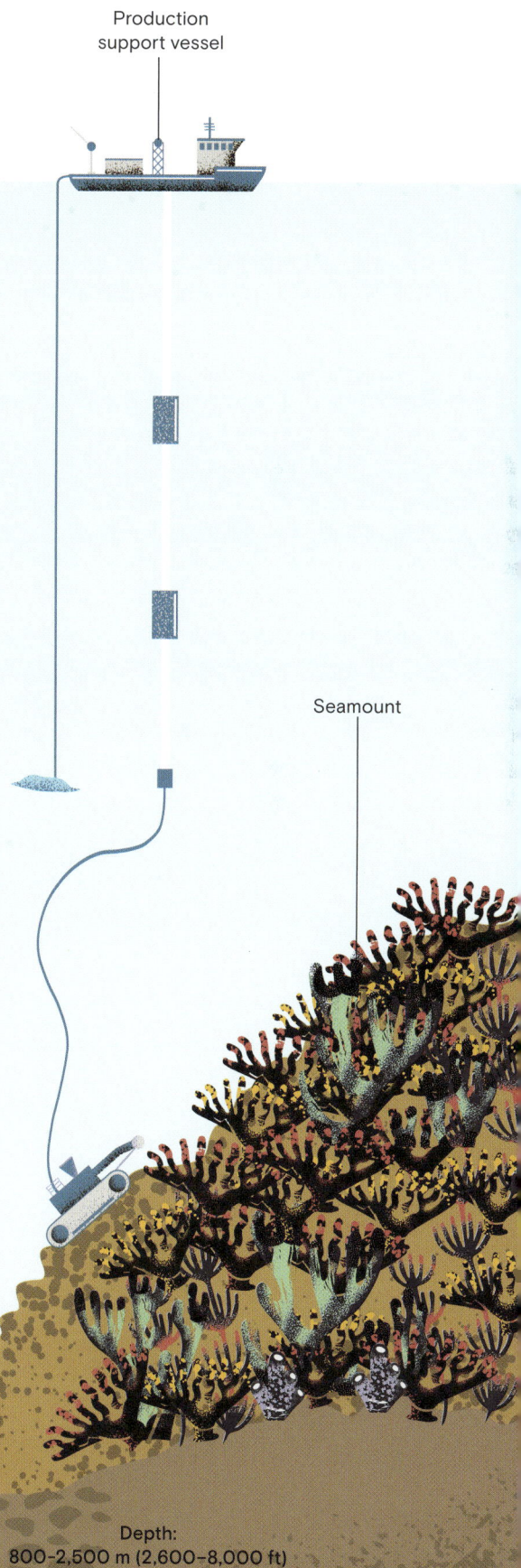

Production
support vessel

Production
support vessel

Lift pump

Vertical riser

Seamount

Seafloor production
tool

Depth:
4,000–6,000 m (13,000–19,700 ft)

Depth:
800–2,500 m (2,600–8,000 ft)

Osedax: The Bone-eating Worm

Deep-sea scientists are notoriously bad at coming up with creative common names for deep-sea creatures. The biggest isopod is a giant isopod. The large, slow shark from the Pacific is the Pacific sleeper shark. The giant worm that lives in a tube is the giant tube worm. But, somehow, deep-sea researchers struck descriptive gold with *Osedax mucofloris*, the bone-eating snot-flower worm, and its bone-eating worm cousins.

Bone-eating worms were discovered in 2002, feasting on the skeletons of long-dead whales. They have since been found throughout the world's oceans. These worms thrive on the carcasses of whales, as well as other nutrient-rich foodfalls including fish, seals and alligators. *Osedax* worms bore into whale bones using specialized rootlike structures that extract nutrients from the hydrocarbon-rich marrow within. This allows them to exploit a food source that is inaccessible to other organisms. The worms play a crucial role in the breakdown of whale carcasses, accelerating decomposition and facilitating the dispersal of nutrients into the deep-sea ecosystem.

Feasting on the bones of fallen whales isn't the only unusual trait of *Osedax*. Female *Osedax* are surrounded by a specialized gelatinous tube, called a lumen. Within the lumen, they host a harem of hundreds of microscopic dwarf males. The females spawn continuously, releasing hundreds of fertilized eggs into the water column to find and colonize new carcasses. This reproductive strategy allows *Osedax* to thrive in places where foodfalls are infrequent events.

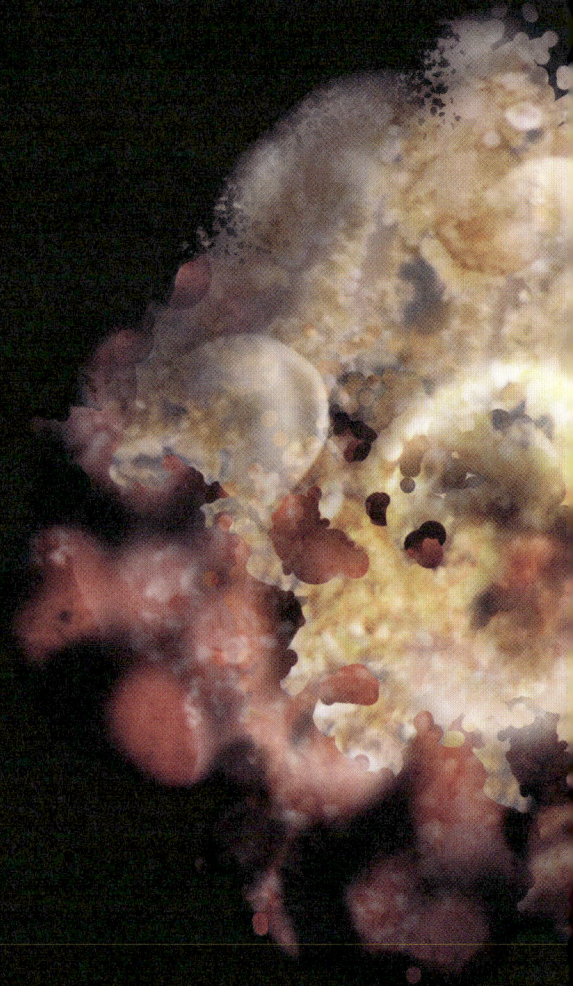

Metals at Sea

Perhaps the most unexpected resource found in the deep sea is both a testament to humankind's desire to explore and a reminder of the legacy of the atomic age: lead. But not just any lead.

In order for physicists to conduct their most sensitive experiments on the building blocks of the universe, they need instruments that can detect individual particles as they travel through the atmosphere. Earth is a very noisy place and these instruments need to be shielded from all the other noise being produced as particles decay. We call that decay radiation, and we shield instruments and other objects against it with lead.

Modern lead mined from the earth naturally contains uranium-235, which decays into lead-210, releasing radioactive particles. The older the lead, the less lead-210 it contains and the less it will interfere with extremely sensitive instruments. The best source of very old lead comes from the remnants of the Roman Empire, and the purest Roman lead comes from shipwrecks, where lead used as ballast was buffered by the sea against exposure to radiation. This lead is the most valuable, as it contains almost no lead-210.

Lead is not the only metal we need to build the most sophisticated and sensitive scientific instruments to understand our physical universe. Much of the modern world is constructed of steel, and modern machines like particle detectors are no exception. There is just one problem. Beginning in the 1940s, humankind developed, tested and detonated nuclear weapons around the world, especially in the Pacific. This spread radioactive particles into the atmosphere and created a clearly defined geologic layer of radioactive particles marking the dawn of the atomic age. The process of refining steel requires atmospheric gas, and all steel manufactured since the end of World War II contains the radioactive legacy of the atomic bomb. Instruments that need to detect radiation, such as Geiger counters and sensors aboard spacecraft, need to be made of steel that is free of that radioactive contamination.

One of the few materials better at blocking radiation than lead is seawater, which is so good at absorbing radiation that it is used to store radioactive waste at nuclear power plants. Steel is not readily found naturally occurring in the deep sea, but there is one unlikely source for it. World War II saw not just the advent of the atomic bomb, but also the acceleration of modern industry. More steel was made to support the war effort than at any prior point in history, and much of that steel, laid into the hulls of warships, found its final resting place in the deep sea.

Low-background steel, produced prior to the first detonation of a nuclear weapon, the Trinity test, in 1945, and buffered from atmospheric exposure by the sea, is an essential component in some of the most delicate scientific instruments. But unlike lead, steel rusts, and as the vast skeletal fleets of global conflict oxidize into oblivion beneath the sea, we are running out of low-background steel.

02

01 – *Aida* shipwreck
Brother Islands, Egypt

02 – Operation Crossroads,
Test Baker, 25 July 1946
The 21-kiloton 'Helen of Bikini' nuclear bomb detonates 27 m (90 ft) underwater in water 55 m (180 ft) deep.

Point Nemo

There is a point in the middle of the Pacific that is further from inhabited land than any other place on the planet. It is so remote and so inaccessible that it became the target for one of the strangest uses of the deep sea. Its name is Point Nemo and it is a graveyard for starships.

The point is named after Captain Nemo in Jules Verne's *Twenty Thousand Leagues Under the Sea*. At roughly 2,700 km (1,670 miles) from the nearest landmass, Point Nemo is so remote that when the International Space Station passes overhead as it orbits the Earth, the astronauts on board are, by a wide margin, the closest humans to the point.

And the International Space Station does pass overhead, several times per day, along with the majority of satellites in orbit. When a large spacecraft, space station or satellite reaches the end of its operational life, it must be deorbited and surrender its journey to Earth's gravity. Small satellites in low-Earth orbit will simply burn up from the heat of re-entry, but larger objects come crashing back into the world. To avoid raining debris upon a populated area, the international community of space-faring nations aims its spacecraft at Point Nemo.

Point Nemo was identified by the USA and the Soviet Union during the zenith of the space race as an acceptable target for deorbiting. Since then, dozens of satellites, several space stations and multiple other orbital assets, some top secret, have made their final descent towards Point Nemo, ensuring that these falling objects make planetfall far from human habitation.

Point Nemo has become the maritime graveyard of the space age. Two hundred and sixty-three spacecraft have been disposed of at Point Nemo since 1971. When the International Space Station is finally decommissioned and deorbited in 2031, it too will be laid to rest at Point Nemo.

Point Nemo, the furthest point from inhabited land on the planet

Oceans & Us

Introduction

'All of us have in our veins the exact same percentage of salt in our blood that exists in the ocean, and, therefore, we have salt in our blood, in our sweat, in our tears. We are tied to the ocean. And when we go back to the sea – whether it is to sail or to watch it – we are going back from whence we came.'

– John F. Kennedy

No matter where we live – on top of a mountain, thousands of kilometres away from a coastline or at the water's edge – our lives are deeply intertwined with the vast expanse of the ocean in ways that we do not and perhaps will never fully understand. Our curiosity has driven us to compulsively imagine, celebrate and revere this watery space through stories and voyages to far-flung places. From *Moby-Dick* to the HMS *Challenger* expedition, *Jaws* to the Kraken, the ocean looms large in the mind's eye.

Meanwhile, just under half our global population depends on fish for at least a part of their protein intake. More than 3 billion people rely directly on the ocean for their livelihoods, with countless others indirectly dependent on this ecosystem. It supplies oxygen, moderates the climate, influences the weather and supports human health. And while we return to the well-worn phrase, 'All roads lead to Rome', it really is the case that all waterways lead to the ocean. The ocean is divided by invisible watery lines, but the truth is that it is one single body of water with visible boundaries only where it laps the shoreline of our countries and continents. The ocean connects us all and is our common heritage; our actions in one corner have ripple effects across this giant tank, so our collective action matters.

For millennia, our dependency on the ocean has driven us to learn how to live with it, in it and around it. To develop relationships with the creatures that call it home. This association has shaped us as a species, sometimes modifying our physiology to enable us to interact better with the ocean. It has pushed our capacity to innovate and create bigger and better ways to travel and transport goods efficiently. It has challenged us to design tools and toys that allow us to go deeper and further. It has dared us to dream of ways to communicate with those species that live beyond our shorelines. But as we have delved deeper into this relationship, humans have become a dominant force of planetary change. We have subsequently started to shape the ocean, resulting in a changed ocean known today as the Anthropocene Ocean.

The ocean connects us all and is our common heritage; our actions in one corner have ripple effects across this giant tank, so our collective action matters.

With every advancement we have made in favour of better, more comfortable lives for ourselves, we have done increasing damage to our oceans. Driving species to extinction or near extinction, increasing noise levels to the point of deafening and disrupting the lives of marine species, and dumping vast quantities of rubbish that will not vanish in our lifetimes but will continuously break down into smaller components and simply linger. We have overfished our oceans, allegedly to keep the world fed, but our inefficiencies and mismanagement have led to the discarding, wastage and inappropriate use

Casting a net to catch fish
Balneário Camboriú, Brazil

The ocean is the largest
ecosystem on our planet.
To protect it, we must build
the world's biggest team.

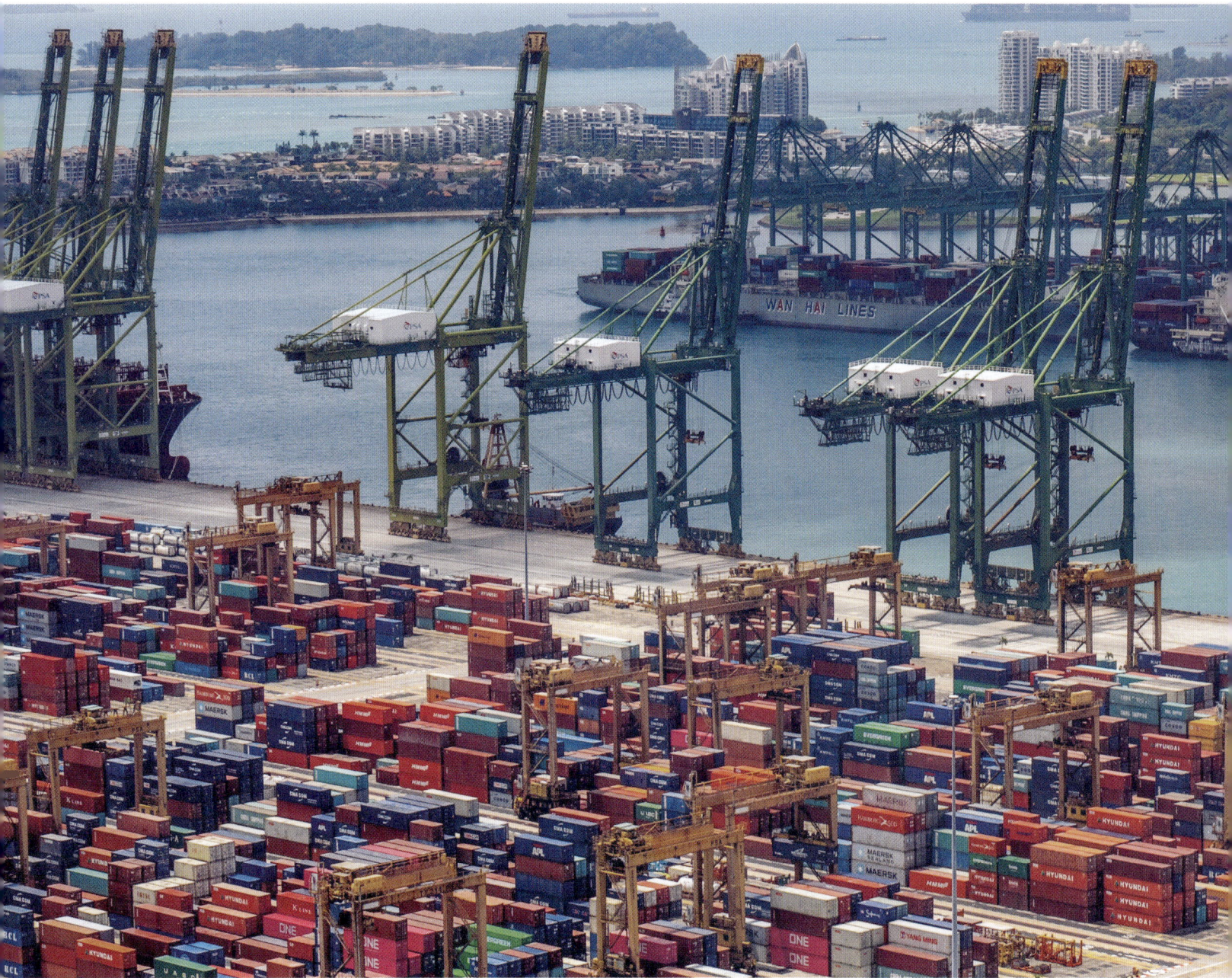

01 – Cranes and shipping
containers at the Port of
Singapore
Keppel Harbour, Singapore

02 – HMS *Challenger* at the naval
base in Bermuda, West Indies
Caleb Newbold (c. 4 April–31
May 1873)

of so many of these valuable stocks. While fishing to feed, we have dumped vast quantities of gear into the ocean, inadvertently or otherwise, with the same result. This discarded, lost or abandoned fishing gear – known as 'ghost gear' – continues to fish our oceans for longer than our lifetimes, entangling and often killing turtles, dolphins and even whales. Every year, we regain the 'hottest ocean temperature on record' title, and we have systematically bid adieu to vast expanses of coral reefs – homes and nursery grounds to countless species. We have disrupted the chemical balance of our oceans and made them more acidic, weakening and dissolving the skeletons and shells of a range of species, even commercially valuable ones. We dredge, mine, explode and build, and as we do so, we modify and degrade the ocean to a point where it cannot support the fauna and flora that call it home. Our actions have irreversibly modified food webs – changing the abundance and distribution of species and leading to cascading effects throughout the ecosystem.

While historically the ocean was seen as an infinite space of extraction and an equally infinite dumping ground, today our understanding of the ocean is better than ever. We now know that nothing is limitless and that our needs and actions can have direct and indirect consequences on even the deepest parts of our planet. With this enhanced understanding and vastly improved knowledge, we must act fast to shift the current trajectory of our oceans. To do this, we must incite and inspire curiosity and make space for more people to tumble into the depths of our ocean as custodians, keepers and protectors.

The ocean is the largest ecosystem on our planet. To protect it, we must build the world's biggest team. To learn from our mistakes, shine a light on the limitless potential of humanity to do good and leave the ocean a better place than we found it.

01

02

A Brief History
of Diving

Humans have been exploring the depths of the oceans for millennia. In ancient times, the Greeks and Romans engaged in freediving to collect food, sponges and pearls. Over time, the first snorkels were fashioned out of hollow reeds, enabling humans to breathe underwater. The ancient Greek Scyllis, captured by the Persians, escaped by swimming nearly 15 km (9 miles) using a makeshift snorkel made from a hollow reed.

As early as the fourth century CE, Aristotle mentioned the use of 'diving bells' to explore the bottom of the ocean. In the thirteenth century, Persian divers made goggles by thinly slicing and polishing tortoise shells. These accounts indicate humanity's curiosity about the ocean and recognition that the resources beneath were extractable and usable.

It was not until 1690 that Edmond Halley, the English scientist who discovered the comet named after him, designed a diving bell. This consisted of two lead-weighted barrels, a large working barrel and a smaller air-refill barrel, with openings on the bottom that allowed air and water to enter. Leather hoses attached to the barrels allowed air to flow to a person working in the bell and a diver working outside in the ocean. In the late 1700s, the refill barrel was eliminated once a method for continuously pumping air into the diving bell was devised.

The original dive suits, from the nineteenth century onwards, were made of leather and accompanied by a heavy copper or brass helmet called a hard hat. Hard-hat divers could remain underwater for long periods by breathing compressed air continuously supplied by a pump at the surface. While this allowed extended time on the bottom, it restricted movement and vision. The technology used today has changed little, albeit with modifications such as the use of watertight rubber 'dry suits' and lighter, smaller helmets or facemasks. This, as well as contemporary insulating rubber wetsuits, has increased manoeuvrability for divers while underwater.

Modern recreational diving came into being with the perfection of the demand regulator, also known as scuba (self-contained underwater breathing apparatus), by Jacques Cousteau and Émile Gagnan as recently as 1942. This device allowed the diver to breathe easily and safely from a tank of compressed air strapped to his or her back and was called the Aqua-Lung.

Much of modern scuba-diving equipment was ultimately developed in response to war. The ability to head to the depths and remain there for extended periods became important during a siege, thus prompting the invention or improvement of both the diving bell and the submarine. These pieces of equipment fulfilled that need while ensuring the safety of the divers involved.

01

01 – Jacques Cousteau, marine explorer and joint inventor of the Aqua-Lung, 1970s

02 – Diving bell with two divers on board, c. 1820
An improved version of Edmond Halley's diving bell

03 – Man wearing a metal diving suit, 1934

04 – Ancient Greek fresco depicting a diver, 6th century BCE
Campania, Italy

05 – Man in a hard-hat deep-sea diving suit, 1930s–40s

06 – Alexander the Great's diving bell, 15th century

O2

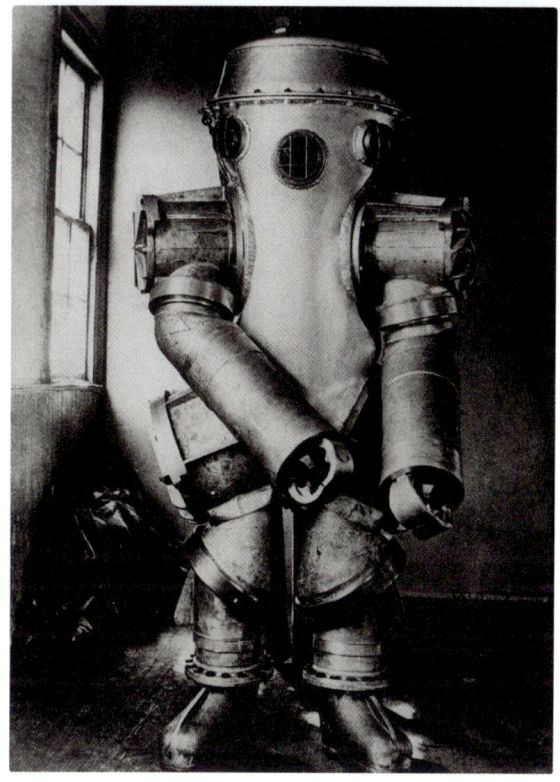

O3

O4

O6

O5

OCEANS & US

Natural-born Freedivers

The Bajau people of Southeast Asia are some of the world's best freedivers. Known as sea nomads, they have an innate ability to dive to depths of over 60 m (200 ft) and stay there for up to ten minutes without wetsuits or fins, using only homemade wooden goggles and weight belts. They walk on the seafloor with the ease and comfort of most people walking on land. They have lived at sea around the Philippines, Indonesia and Malaysia for more than 1,000 years on small 'Lansa' houseboats.

Their extreme diving skills are possible because of a distinctly enlarged spleen – an underrated organ that filters your blood day in and day out. While marine mammals have notably large spleens compared to their land-based counterparts, the Bajau, who have spent thousands of years on the ocean, have spleens that are 50 per cent larger than regular land-living humans. Since each contraction of your spleen squeezes blood into the circulatory system, adding more oxygen into the mix, having a bigger spleen means even more oxygen is available with every contraction.

Unfortunately, the unique way of life of these sea nomads, traditionally built on fishing and freediving, is currently under threat. Destructive fishing methods such as dynamite and cyanide fishing, together with warming ocean temperatures affecting their hunting grounds and consequently threatening their traditional way of life, have forced at least some of them to move to dry land over the last few decades.

01

02

01 – A Bajau freediver swimming among a shoal of jackfish
Tun Sakaran Marine Park, Sabah, Malaysia

02 – Bajau children catching an octopus
Semporna, Sabah, Malaysia

Haenyeo and Ama

The term *haenyeo* comprises two parts: *hae* means sea, and *nyeo* means women. *Haenyeo*, from the island of Jeju off southern Korea, are literal sea-women. Each is equipped with a *tewak*, a collection net attached to a large orange float. These *tewak* and floats go overboard first to mark each diver's patch for the day, followed closely by their owners. Nowadays, *haenyeo* are mostly grandmothers aged between sixty-five and seventy-five – women who have engaged in this profession for decades. While in the earliest days they wrapped their bodies in white cotton, by the 1970s things had changed, with modern *haenyeo* clad in neoprene wetsuits for the long day ahead. Once in the water, the freedivers hyperventilate, attempting to saturate their blood with oxygen before diving down to the seabed, where they search for marine species – from shellfish to octopuses, conches to sea urchins – which they will later sell at the market. For up to ninety days of the year, these sea-women dive down to 10 m (33 ft) for up to seven hours a day, holding their breath for a minute each dive. On return to the surface, they inhale and exhale, making a unique sound with their breath called *sumbisori*, a rapid panting that sounds like a whistle.

Haenyeo are traditionally community-oriented, performing rituals before their day on the sea together and coming together at the end in huts (often a stone-walled *bulteok*) to be warmed by a fire. This profession, more than four centuries old, embraces sustainable practices. For example, women take only shellfish that have grown over 7 cm (2¾ in) in size – anything smaller is returned to the water to grow. With abalone on the decline, products such as turban snails are collected under strict guidelines from the sea-women and their representative fishing associations.

Diving repeatedly in water as cold as 3°C (37.4°F) during winter is no easy task, but their passion for the ocean trumps their desire to shy away from the challenges they face. It is no wonder that the tradition was added to the UNESCO Intangible Cultural Heritage list in 2016.

In neighbouring Japan, another group of women have embraced the sea and have been freediving to harvest seafood from coastal seabeds for perhaps as long as 2,000 years, and as such thought to be a considerably older tradition than *haenyeo*. *Ama* divers, also known as the women of the sea, historically dived nude, with only a loincloth (later trunks) to hold a knife and a headscarf to keep their hair back. Over time, they transitioned to wearing white costumes to protect themselves from the cold and also, apparently, from sharks. The *ama* divers were particularly instrumental in establishing Mikimoto Pearl Island, where they would dive to plant and harvest pearl oysters on the seabed.

Unfortunately, these traditions are slowly vanishing in both countries, with *ama* and *haenyeo* becoming older and fewer young women willing to take their place.

01

01 – *Ama* divers preparing for diving on a pebble beach
Mie, Japan

02 – *Haenyeo* underwater shot
Jeju Island, South Korea

03 – A female diver preparing to collect pearls
Mikimoto Pearl Island, Japan

02

03

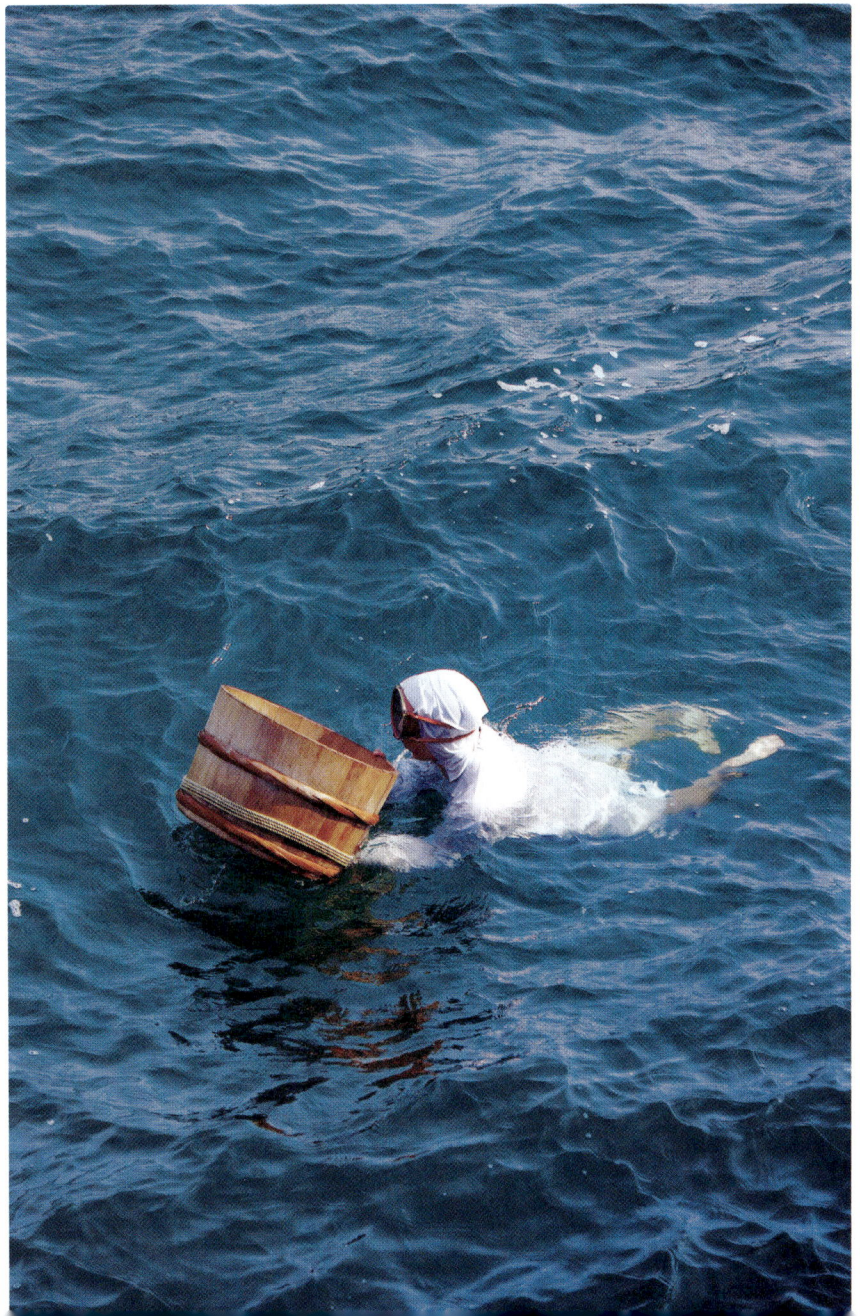

01 – Illustration of marine creature from *Historia animalium*, vol. 4
Conrad Gesner (1558)

02 – A Nereid riding a sea centaur accompanied by other sea creatures
Angiolo Falconetto (c. 1540–60)

03 – 'Mermaids exhibited successively in the years 1758, 1775, & 1794'
Vintage engraving for the *Encyclopaedia Londinensis* (1817)

04 – *La Sirène* (*The Siren*)
Photogravure print from the original painting by Charles Landelle (1883)

05 – Colossal octopus or kraken attacking a sailing vessel
Illustration by Pierre Denys de Montfort (c. 1802)

01

MERMAID.

Mermaids exhibited successively in the Years 1758, 1775, & 1794.

03

02

04

Mythical Marine Creatures

Hundreds of years ago, Scandinavian sailors would talk about a legendary but aggressive sea monster with a taste for flesh and the capacity to sink ships, taking sailors to their doom. This giant monster, the Kraken, struck fear into the hearts of sailors travelling between Norway and Iceland. Today, we believe that these sea-beast legends were based on giant squid of the genus *Architeuthis*, which had a diet of crustaceans and fish but certainly not ships. While the females grow to sizes of 13 m (42½ ft) and males to 10 m (33 ft), sinking a ship would be quite a feat. The rarity of their sightings doubtless added fuel to the fire of the sailors' imagination, but in reality, if they are seen at the ocean's surface, they are typically stressed and dying because they cannot get adequate amounts of oxygen.

As with Kraken, mermaids have also appeared in folklore and myth over the centuries. One of the oldest references to a mermaid, also known as a siren, a half-human, half-fish creature (although sometimes half-bird), comes from ancient Greek mythology. In Homer's *Odyssey*, the sailors are lured by the sirens' beautiful melody, which causes them to turn into shallow waters and crash on the rocks. Much later, when Columbus was exploring the Caribbean in the fifteenth century, he had a 'mermaid' sighting of his own. Today, this sighting is regarded as the first written record of manatees in North America. This confusion between the blubbery marine mammal that is a manatee and the beautiful half-woman half-fish that is a mermaid seems absurd today, but given how little was known about the ocean and its inhabitants at the time, it is excusable.

05

The Humboldt Current and Peruvian Guano

Ocean currents are important in distributing heat across our planet and regulating and stabilizing climate patterns. Surface ocean currents are wind-driven, resulting in horizontal and vertical water movement. Horizontal surface currents are short-lived and include rip currents as well as longshore currents. Upwelling currents move water vertically and are responsible for bringing cold, nutrient-rich, dense water from the depths of the ocean to the surface, pushing warmer, less dense water downward, where it condenses and sinks. This cycle of upwelling and downwelling has important consequences.

The Humboldt Current, a low-salinity current that starts in the frigid waters of Antarctica and runs north along the western coast of South America up to Peru, is one of the most crucial ocean currents in the world. Named after the pioneering German naturalist Alexander von Humboldt, who conducted groundbreaking research in the region in the eighteenth and nineteenth centuries, the resulting upwelling leads to water colder than expected at an equatorial latitude (14–24°C vs 28°C, or 57–75°F vs 82.4°F). It has been the main driver of one of the world's most productive commercial fisheries since the mid-twentieth century, comprising anchovy, sardines and mackerel. Roughly 15 per cent of global fish catch, amounting to about 9 million tonnes of fish and other seafood, is caught because of the Humboldt Current running between Chile and Peru annually. The anchovy fishery represents the largest single-species fishery in the world.

The productive waters also support a wide diversity of other species, including birds – such as the Peruvian pelican, Peruvian booby and guanay cormorant – that partake in the fishy feast. For thousands of years, millions of these seabirds would gorge on anchovies before returning to barren, rocky and vegetation-less islands in the vicinity, where they would defecate. As a result, the droppings would accumulate and bake on these islands, and because of the arid climate and minimal rain resulting from the large swathe of cold water brought to the surface by the Humboldt Current, the valuable nitrates in the guano (the name for the accumulated excrement of seabirds) were preserved. At the peak of the guano boom, approximately 60 million birds producing unadulterated droppings attracted the attention of humans, who promptly began to use the guano as a fertilizer for their crops. In the nineteenth century, Europeans recognized the value of this 'white gold' and introduced it into world trade. From 1830 to the early 1870s, bird droppings were single-handedly responsible for boosting the Peruvian economy.

Today, the population of these guano birds is in decline, with numbers estimated to be around 5 million. Recognizing their importance – ecologically, biologically and economically – Peruvian authorities are trying to sustainably manage the guano industry by protecting the habitat of these seabirds. Unfortunately, beyond this, the seabird populations face threats from the overfishing of their prey, the anchovy stocks, particularly for the fishmeal industry.

01

02

NEGATIVE BY H. MOULTON.   Entered, according to Act of Congress, in the year 1865, by Alex. Gardner, in the Clerk's Office of the District Court of the District of Columbia.   POSITIVE BY A. GARDNER.

Rays of Sunlight from South America.

THE GREAT HEAP—2,000,000 TONS GUANO—CHINCHA ISLANDS.

03

Cooperative Fishing

While dolphins are often seen as competitors to fishers, one small town in Brazil called Laguna has a different story to tell. For more than 140 years, starting in the nineteenth century, fishermen and dolphins in this area have engaged in cooperative fishing, resulting in mutual benefits for everyone – except the fish.

Resident Lahille's bottlenose dolphins work to herd the area's mullet, an important food source for local people, towards a line of fishers standing knee-deep in water along the coast. Once the fish have been compacted, the dolphins signal with head or tail slaps or sudden deep dives in front of the fishers, signs that the fishers have learned to interpret to indicate where and when they should cast their nets. By working in synchrony, the dolphins and traditional net-casting fishers catch more fish than they would if they worked independently, resulting in very little conflict between the species.

This form of cooperative fishing has existed for more than a century and is passed on through social learning, but only about twenty dolphins consistently engage in this activity. Essentially, this single pod of dolphins supports some 200 fishers with no other source of income, highlighting this relationship's important economic and social value. While nothing in the environment stops other dolphins from participating, those that rely on alternative food sources tend to use areas that overlap with other fisheries' activities, increasing the dolphin bycatch rate.

Fishermen work with dolphins to corral their catch

01

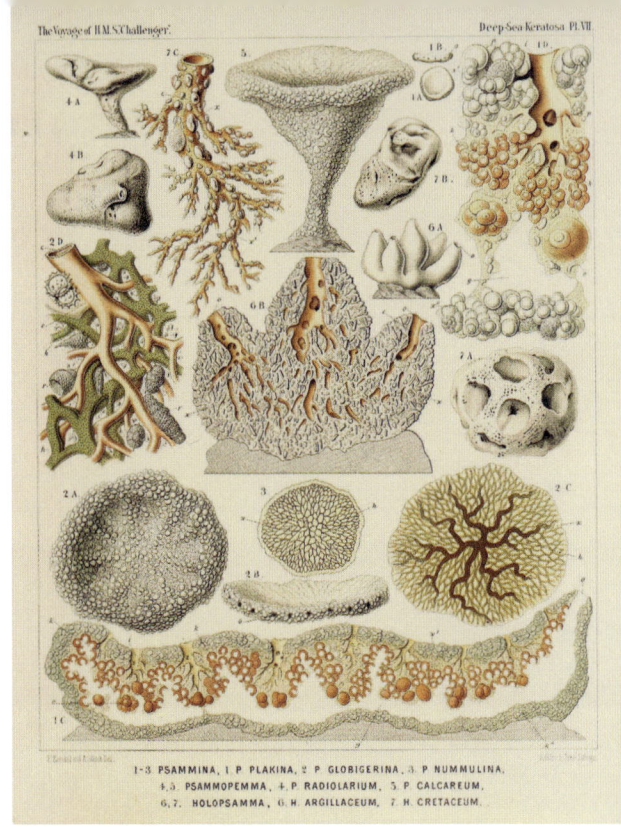

02

03

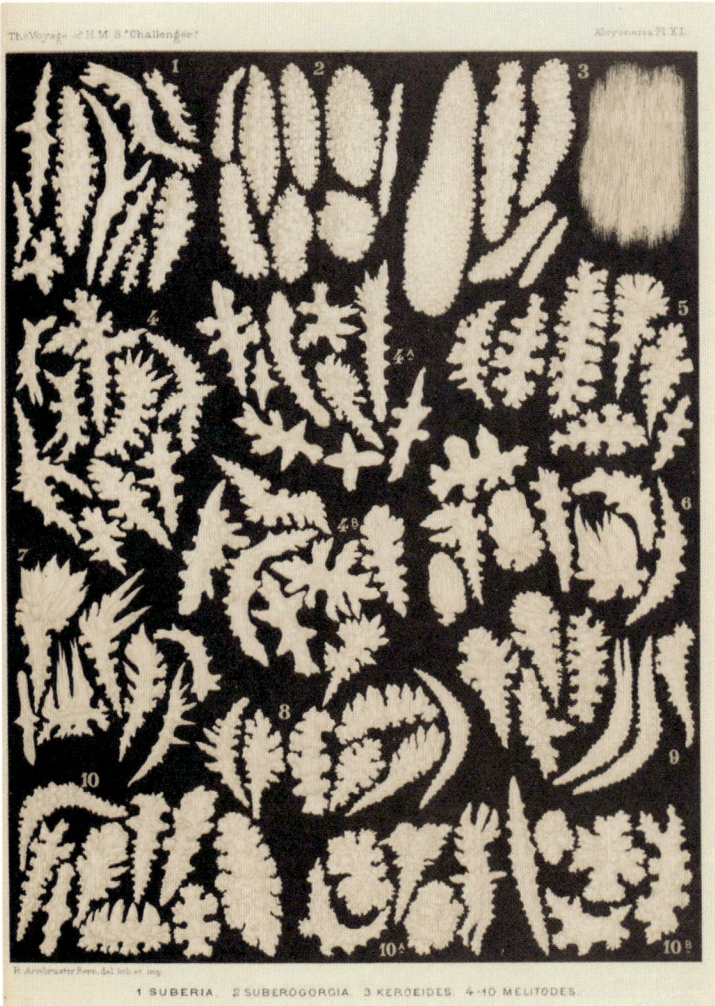

04

01 – Scientific staff and crew members on deck of HMS *Challenger* (1858)

02, 03, 04 and 06 – Reports of the scientific results of the HMS *Challenger* mission

05 – HMS *Challenger* preparing to set sail

HMS *Challenger*

05

In the late nineteenth century, on what is often considered the first true oceanographic expedition, HMS *Challenger* circumnavigated the globe, collecting water, sediment and sea-life samples at various depths. The voyage, lasting 1,000 days and covering more than 126,000 km (68,000 nautical miles), carried out observations at 362 deep-sea stations. Sponsored by the British government, the expedition was like none other, as it laid the foundation for modern oceanography and greatly expanded our knowledge of ocean currents, marine biodiversity and the seafloor.

The voyage was largely possible because HMS *Challenger* was well equipped to explore deeper than ever before. Scientists on board used rope and hemp to lower trawls, nets and samplers to the ocean floor to collect specimens, which were stored in jars and peered at through microscopes. Every few days, they used sounds or would drop a weighted line to measure the ocean's depth. Special thermometers were employed to measure the temperature of the water at different depths. They also collected seawater samples to determine the salinity of the ocean at the different stations. The resulting report, titled *Report of the Scientific Results of the Exploring Voyage of HMS Challenger during the years 1873–76*, was 29,500-pages long, ran into fifty volumes and took twenty-three years to compile. It disproved the notion that life did not exist below 5,500 m (1,800 ft); recorded 4,717 new species; provided the first systematic plot of currents and temperatures in the ocean; included a map of bottom deposits; and provided an outline of the main contours of the ocean basins, including a rise in the middle of the Atlantic Ocean, known today as the mid-Atlantic Ridge (see pp. 210–11). During a depth sounding on this expedition, the team on board made a landmark discovery – the Mariana Trench, the deepest point on Earth (see pp. 218–19).

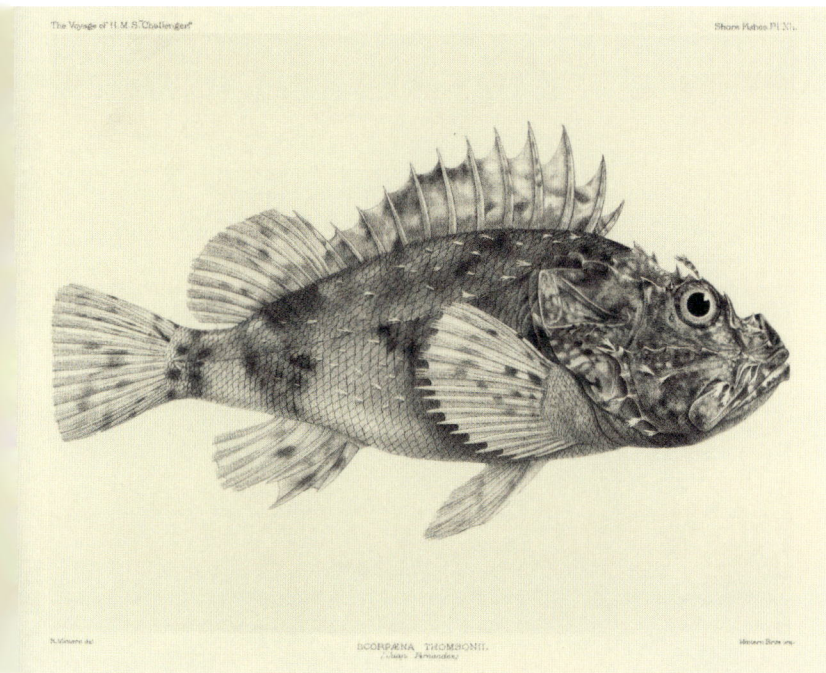

06

Polynesian Wayfinding

Starting around 1500 BCE – long before the invention of sextant, compass and GPS – voyagers seeking to reach new lands navigated the vast expanse of the Pacific Ocean, greater in size than all terrestrial landmasses combined, using the sun, moon, stars, planets, winds and ocean currents. Stories of the voyages of Polynesian ocean-going travellers are evidenced in petroglyphs, by the observations of European explorers and through the tales passed down through Polynesian oral tradition.

Successful voyages required a number of key ingredients, including well-built canoes, skilled navigators and favourable weather conditions. The double-hulled canoes, or *wa'a kaulua*, powered by sails and steered by a single oar and filled with all their supplies, domesticated animals and planting materials essential to start life in a new place, were built by the community, including canoe builders, navigators, priests and hula dancers. Navigators used their powers of observation of the natural world to identify the direction of tradewind-generated ocean swells by watching the rocking motion of their boats. Sunrise and sunset helped them to tell the difference between east and west. The star compass required them to memorize the rising and setting points of stars and constellations at different times of the year. They then divided the horizon into thirty-two 'houses', with the canoe in the middle. This way, they could recognize directions by watching the rising and setting of various stars. The North Star, the only fixed point in the sky as the Earth rotates, always indicates north; its presence helped navigators in the northern hemisphere, and its distinct absence in the southern hemisphere led navigators to turn to the Southern Cross for directional help.

Clouds provide cues about the weather but also indicate potential landmasses. High masses of clouds could signal mountainous islands. As the sailors got closer to land, they would rely on known species of fish and seabirds and floating objects such as debris or vegetation to determine proximity to land. Using these techniques, Polynesian wayfinders settled more than 1,000 scattered islands across the Polynesian Triangle between New Zealand, Hawaii and Easter Island. The ocean and its inhabitants have supported humans in more ways than we fully understand.

01 – *Pahi of the Tuamotu Islands*
Herb Kawainui Kāne
(1928–2011)

02 – Old Polynesian navigation device showing directions of winds, waves and islands

01

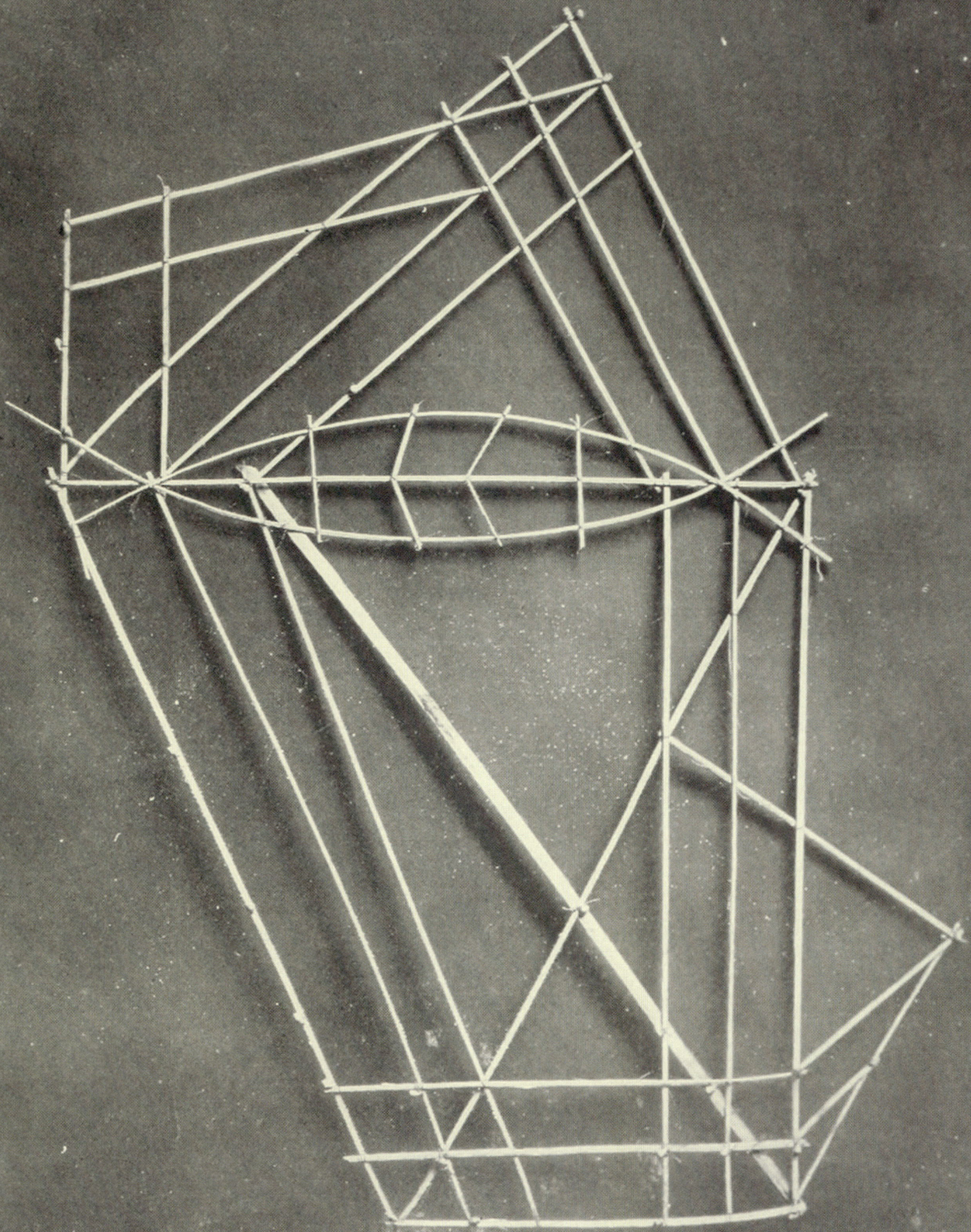

Maritime Trade

Maritime trade has been integral to human civilization for thousands of years, shaping global economies, cultures and geopolitics. The ancient maritime trade routes that crisscrossed the oceans facilitated the exchange of goods and enabled the mingling of cultures, ideas, technologies and cuisines, turning our world into what it is today. From spices to ceramics, the ocean was the main means of transporting goods and a means to connect disparate areas of our world.

The earliest records of maritime trade date back to 3000 BCE and to ancient civilizations such as the Phoenicians and the Egyptians, which used waterways to trade grain, textiles, pottery and precious metals. With the rise of the Byzantine Empire and the Chinese dynasties, maritime trade opened up across the Mediterranean, Indian Ocean and Silk Road routes. The Age of Exploration (fifteenth to seventeenth centuries) led to the existing European powers, including Portugal, the Netherlands, England, Spain and France, embarking on expeditions to find new trade routes to Asia. This was also the age of the establishment of colonial empires, which led to the exchange of major commodities, between continents, including spices, silk, tea, coffee, sugar and enslaved people. The more recent Industrial Revolution was a time of great transformation, with advancements in shipbuilding, navigation and manufacturing setting the stage for the advances of the twentieth century, which saw significant advances in maritime technology.

Today, 80 per cent of merchandise is transported using maritime routes. Shipping is still the most efficient means of transporting goods from one place to the other. The world's merchant fleet increased fifty-fold between 1995 and 2020, and expectations are that maritime traffic will continue to increase alongside global economic growth.

While trade across the ocean is unavoidable in an increasingly globalized world, its impacts on the ocean and its inhabitants are massive. In areas where shipping routes overlap with megafauna that depend on surface waters for survival, such as whales, turtles and whale sharks, ship-strikes are inevitable and are fast becoming one of the biggest threats to these populations. Additionally, before the Industrial Revolution of the nineteenth century, low-frequency sound in the ocean was largely generated by spray and bubbles associated with breaking waves, noises generated by marine life and other natural sounds. The Industrial Revolution represents a key point in maritime history as it marked the beginning of powered (rather than sailing) vessels to transport goods and provide services. As a result, low-frequency noise levels have gradually increased in the ocean since the Industrial Revolution and today represent the most pervasive anthropogenic noise. In a world where most species depend on sound as their primary sense, these anthropogenically produced sounds and increased noise levels are disruptive, with impacts ranging from masking, where noise interferes with an animal's ability to perceive a sound, to permanent or temporary hearing loss, behavioural changes and even strandings.

01

03 – Aerial view of a cargo ship carrying containers

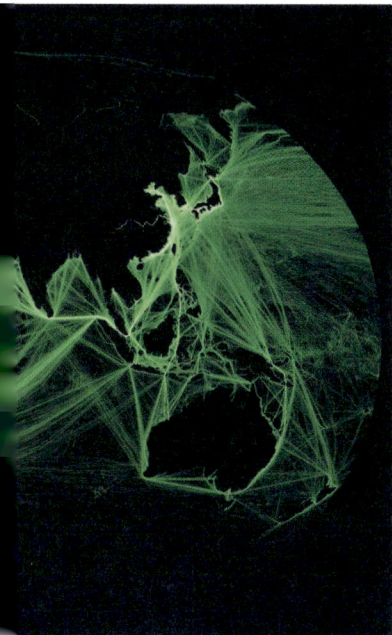
02

An influx of river water can raise or lower the amount of oxygen at a particular spot. Excessive amounts of nutrients making their way into the ocean from land results in eutrophication. This in turn simulates the rapid growth of aquatic plants such as algae. When this algal bloom dies out, the process of decomposition depletes the oxygen levels in the area.

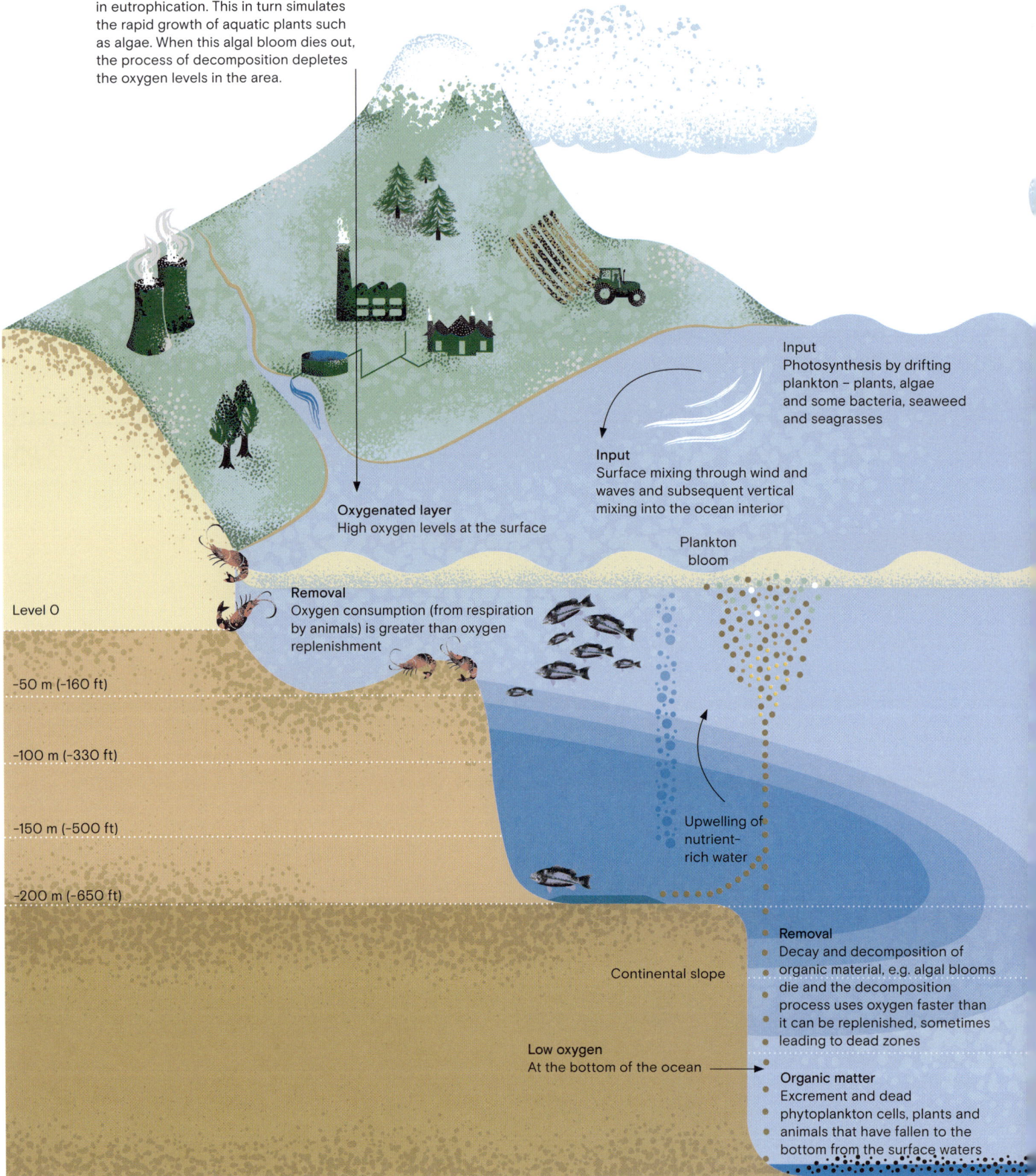

Input
Photosynthesis by drifting plankton – plants, algae and some bacteria, seaweed and seagrasses

Input
Surface mixing through wind and waves and subsequent vertical mixing into the ocean interior

Plankton bloom

Oxygenated layer
High oxygen levels at the surface

Removal
Oxygen consumption (from respiration by animals) is greater than oxygen replenishment

Level 0

−50 m (−160 ft)

−100 m (−330 ft)

−150 m (−500 ft)

−200 m (−650 ft)

Upwelling of nutrient-rich water

Continental slope

Removal
Decay and decomposition of organic material, e.g. algal blooms die and the decomposition process uses oxygen faster than it can be replenished, sometimes leading to dead zones

Low oxygen
At the bottom of the ocean

Organic matter
Excrement and dead phytoplankton cells, plants and animals that have fallen to the bottom from the surface waters

COAST

5–100 km (3–60 miles)

up to 3,000 km (2,000 miles)

Oxygen in our Ocean

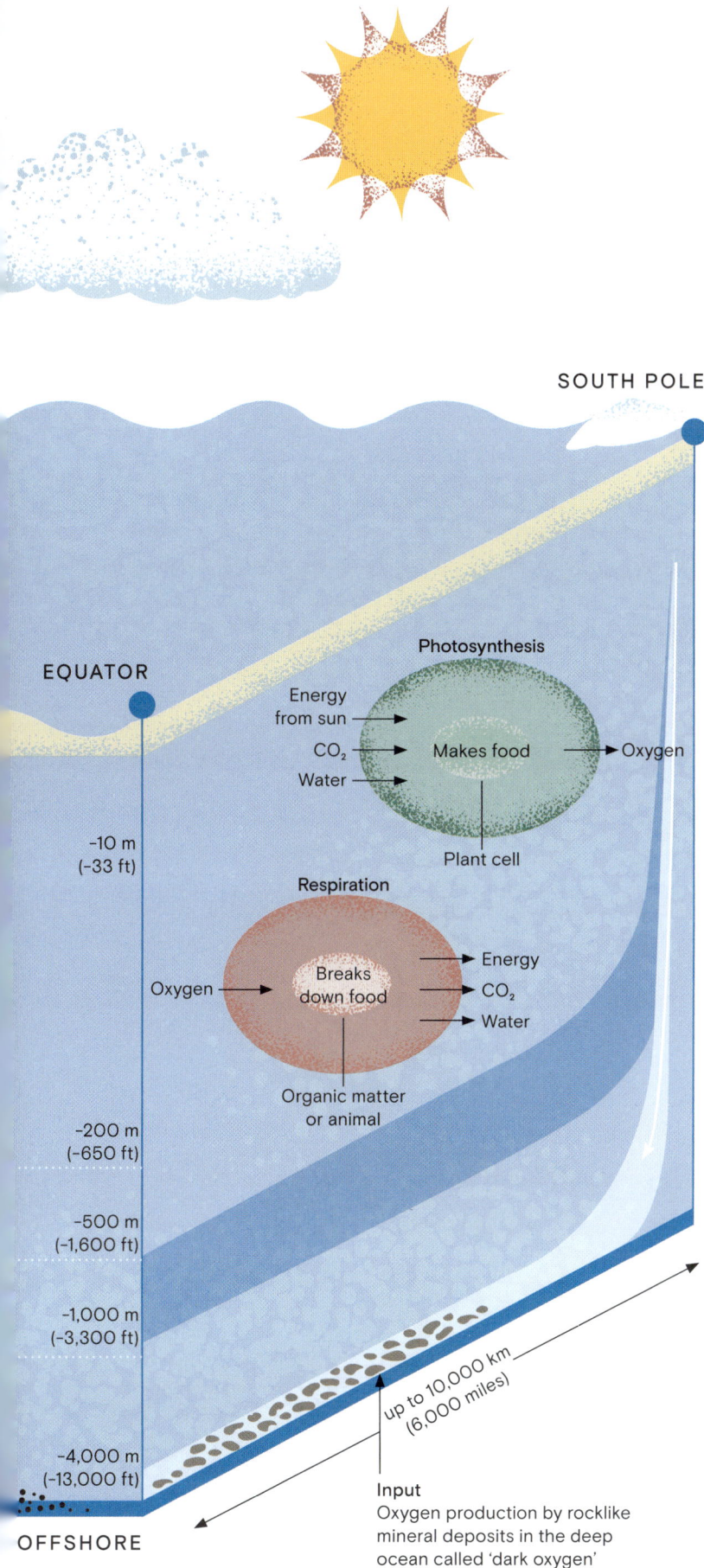

SOUTH POLE

Photosynthesis

EQUATOR

Energy
from sun →

CO₂ →   Makes food   → Oxygen

Water →

Plant cell

-10 m
(-33 ft)

Respiration

Oxygen →   Breaks
down food   → Energy

→ CO₂

→ Water

Organic matter
or animal

-200 m
(-650 ft)

-500 m
(-1,600 ft)

-1,000 m
(-3,300 ft)

up to 10,000 km
(6,000 miles)

-4,000 m
(-13,000 ft)

Input
Oxygen production by rocklike
mineral deposits in the deep
ocean called 'dark oxygen'

OFFSHORE

Half of the oxygen produced on Earth comes from the ocean, mostly produced by oceanic plankton (see pp. 104–5) – drifting plants, algae and some photosynthesizing bacteria. Photosynthesis requires these plants to take up carbon dioxide from the atmosphere and produce oxygen in the top 100 m (330 ft) of the ocean owing to the process's dependence on light. *Prochlorococcus*, the smallest photosynthetic organism on Earth, produces 20 per cent of the oxygen in our entire biosphere. This alone is higher than is produced by all tropical rainforests combined. However, most of what is produced is used within the ocean by marine life – for respiration or during decay.

The colder and saltier the water is, the more oxygen it holds. Because oxygen dissolves in cold water, the colder the water, the more oxygen it can dissolve. However, as the climate changes and the seas continue to get warmer, the surface layer of water becomes warmer and does not mix with the deeper layers below. This means that the oxygen produced by phytoplankton in the surface layers cannot get into deeper waters. This can result in an 'oxygen minimum zone', where little or no oxygen is available. Deprived of sufficient oxygen, particularly in tropical seas, the marine life in these areas becomes stressed, with reproduction, growth and resilience to disease all adversely affected; this can impact fish stocks and, ultimately, the humans who depend on fish as their source of protein or for their livelihoods.

Warming Oceans and Marine Megafauna

No conversation about the ocean is complete without considering climate change and its impacts on the ocean. Every year, humanity outdoes itself by pushing sea surface temperatures to new highs. We continue to gasp as we record the hottest temperatures in history and do not seem to be slowing down.

As environments change thanks to warming oceans, resident species have few options. They can adapt to the increasing temperature, learn to tolerate it, or move – if not, they risk extinction. If the temperature increase happened gradually, adaptation and increasing tolerance would be more feasible. However, the world we live in today is warming at an alarming rate, and this increase in temperature is so sudden that gradual adaptation or even tolerance is rarely possible. The option of moving to avoid rising temperatures is a real one, and in the ocean we expect species to move to deeper depths or higher latitudes (as the oceans are warming polewards). This means species from the tropics are predicted to move towards polar and temperate waters.

But movement is not straightforward. Not only is it possible that species will move into unknown habitats, bringing them face to face with species that already live there – leading to potential competition – but these marine species may move into new habitats in which they will face new and perhaps even bigger threats. In some cases, warming temperatures might make habitats more conducive to anthropogenic activities that did not occur previously, further increasing threats to resident species. For example, shipping activity has expanded as sea ice melts in the Arctic. This leads to an increased risk of collision between bowhead whales and ships (known as ship-strike) and increased noise pollution for this species.

The situation is also dire for species like coral that cannot move. When ocean temperatures increase rapidly, corals undergo a process called bleaching. As the name suggests, during bleaching, the corals lose their vibrant colours owing to the expulsion of symbiotic algae called zooxanthellae that live within their tissue (see pp. 108–9). While corals can survive short-term bleaching events, prolonged or severe bleaching can harm coral health and reef ecosystems. When corals expel these algae, their transparent tissues reveal the white calcium carbonate skeleton underneath, giving them a bleached appearance. This colourless coral skeleton is increasingly vulnerable to external stressors and wholly dependent on external food sources. If the bleaching event is prolonged, the reef system will deteriorate, leading to a degraded reef that cannot provide ecosystem services, such as nursery grounds for many fish species.

Seagrass Beds

-Increasing sea-surface temperatures affect seagrass growth rates.
-Redistribution of seagrasses is caused by rising seas, increased seawater temperature and salinity, and changes to freshwater systems.
-Reduction in plant productivity occurs as a result of increased water depth, limiting the amount of light, water motion and tidal circulation.

Salmon

-Redistribution of salmon species occurs as cold-water habitats become harder to find.
-Warmer water contains less oxygen, making it harder for salmon to breathe.
-Disruption of the wider oceanic ecosystem is seen as climate change influences productivity at the base of the food web.

Impacts on Biodiversity

Mangroves

-Redistribution of mangrove habitats is caused by increases in temperature and rising seas.
-Previously sheltered by coral reefs, mangrove forests are damaged and lost, pummelled by wave action and strong winds.

Marine Turtles

-Sea-level rise and beach erosion reduce the number of nesting habitats.
-Changes occur in sand temperature, which plays a critical role in defining sea-turtle sex. Higher sand temperatures favour female hatchlings, upsetting male–female ratios and thus compromising species survival.
-Coral bleaching and seagrass mortality reduce foraging sites and prey availability.

Coral Reefs

-Increasing sea-surface temperatures lead to coral bleaching and mortality.
-Weakening skeleton and reduced growth rate of calcium carbonate skeleton caused by ocean acidification sees coral loss.
-Degradation of reefs is caused by an increase in the severity and frequency of storms and hurricanes.

Sharks

-Lack of food sources forces changes to geographical species distribution and migration patterns, increasing their interactions with humans.
-Degradation and loss of mating, nursery and foraging areas (mangroves, seagrasses and coral reefs) that are critical for sharks' survival and development are seen.

Ocean-inspired Innovations

Beyond being integral to the health of our planet, the ocean contributes to humanity in other ways. Some ocean inhabitants have inspired innovations that enable us to build state-of-the-art technology more efficiently. While there are many examples, four unique technologies that have benefited from inspiration from our underwater friends are adhesives; strong lightweight materials; antifouling paint; and wind-power efficiency.

Mussels are sedentary, clinging to a single rock and to each other for dear life in the face of hungry predators and crashing waves. They do this by creating 50–100 byssal threads – strong, silky fibres made from proteins, known as the mussel's 'beard' – and then using their sensitive foot to mould the threads and apply a waterproof adhesive to attach to rocks. By studying this process and working out that the glue gains its waterproof properties from an amino acid called Dopa, scientists have been able to create one of the strongest synthetic adhesives available. The mussel-inspired adhesive is created using a biomimetic polymer model that contains Dopa,

and performs ten times better than commercial adhesives when used to bond polished aluminium. It is hoped that this innovation will be utilized for delicate surgeries or to heal wounds.

Mantis shrimp, meanwhile, are known for their strong yet light exoskeletons. They gain this strength because of a unique herringbone structure, which allows the mantis shrimp to protect itself when delivering a blow at speeds faster than a .22 calibre bullet, and enables it to inflict significant damage on its prey. Scientists inspired by this tough, lightweight material are now investigating this structure with a view to using it in composite materials that need to be lighter and stronger with improved impact resistance, such as body armour, football helmets and even aeroplanes. Because less material is used in its production, it is also cost-effective.

Antifouling technologies owe their debt to sharks, which are rough to the touch thanks to the denticles (tiny toothlike scales) on their skin that reduce drag and friction. These dermal denticles serve a valuable purpose as they inhibit the growth of bacteria, ensuring that the sharks stay clean.

When a similar technology called Sharklet is introduced to ships and submarines, it optimizes the vessel's performance, reducing drag and inefficiencies caused by fouling (the build-up of barnacles, algae, seaweed, etc.) and thereby reducing fuel use too.

Last but certainly not least, we turn to one of the giants in our oceans: humpback whales. The tubercles, or small bumps, on their long flippers reduce drag and allow them to be extremely agile in the water. By replicating the tubercles on the blades of wind turbines, researchers have shown a 32 per cent reduction in drag, increasing efficiency and power generation.

03

02

02 – Turbine blades inspired by humpback whale tubercles

01 – Tubercles on a humpback whale flipper increase agility and reduce drag

03 – Sedentary zebra mussels, which inspired the creation of the strongest synthetic adhesives available

Glossary

aggregation
A gathering of fish of the same species for spawning – releasing eggs and sperm into the water for fertilization, leading to reproduction.

anadromous
Applied to fish species that spend portions of their lives in both salt and fresh water, migrating from the ocean to rivers and streams to spawn.

archaea
Single-celled microorganisms that constitute one of the three domains of life (Archaea, Bacteria and Eukarya) but remain evolutionarily distinct. Archaea live in low-oxygen, often extreme, environments.

barbel (chin barbel)
A sensory organ located on the chin or lower jaw of some fish species. It contains sensory cells that help fish detect food and navigate their environment.

barrier island
A barrier island runs parallel to the mainland coast and is often long, narrow and made of sand or sediment.

bathypelagic
A deep-sea zone that extends between 1,000 and 4,000 m (3,300 and 13,000 ft) below the surface.

bioluminescence
The production and emission of light by living organisms such as deep-sea fish.

bivalve
Marine molluscs – soft-bodied invertebrates, such as clams and oysters – that have a two-part shell joined by a hinge.

brackish
Brackish water is more saline than fresh water but less saline than the ocean.

calcareous
Containing calcium carbonate.

caldera
A large depression that is formed shortly after a volcanic eruption.

cartilaginous
Having a skeleton entirely or predominantly made up of cartilage. Cartilaginous fish include sharks and rays.

cephalopod
A member of the molluscan class Cephalapoda, including squid and octopus. From the Greek meaning 'head foot'.

chemosynthesis
The process by which organisms produce food using chemical energy.

chitin
A natural biopolymer, chitin is a horny polysaccharide that is a major constituent of the exoskeleton of insects.

chiton
An ancient marine mollusc that is flattened, comprises eight plates and is bilaterally symmetrical.

chordate
Animals that belong to the phylum Chordata, which consists of animals with a flexible rod supporting their dorsal (back) sides. This encompasses vertebrates, including fish, amphibians, reptiles, birds and mammals.

colonial organism
A collective life form composed of many physically connected, interdependent individuals.

continental rise
The sloping transition zone of accumulated sediments located between a continental slope and abyssal plain.

copepods
A group of small ubiquitous aquatic crustaceans.

countershading
A form of camouflage or cryptic coloration where the upper surface of an animal's body is dark and the underside is light to enable blending into the surroundings.

cyanobacteria
Free-living single-celled photosynthetic organisms informally referred to as blue-green algae.

denticle
Any small toothlike or bristle-like structure.

diatoms
Single-celled algae that can be solitary or live in colonies.

dinoflagellates
Single-celled organisms bearing two dissimilar flagella and having characteristics of both plants and animals.

dorsal
Belonging to or on or near the back or upper side of the body.

ecotype
A population that is genotypically adapted to living in a particular environment.

epipelagic
The topmost zone of the oceans, between 0 and 100 m (0 and 330 ft), where sunlight is present and photosynthesis occurs.

euryhaline
An aquatic animal that can tolerate a wide range of salinities.

fathom
A unit measure of water depth equal to 6 ft (1.8 m).

filter feeding
A foraging strategy in which animals feed by straining food particles or small organisms from the water.

fusiform
Spindle-shaped.

gastropod
Belonging to the phylum Mollusca, gastropods are characterized by having a muscular foot, eyes, tentacles and a rasp-like feeding organ called the radula, comprised of many tiny teeth.

hadal zone / hadal trenches
The deepest part of our ocean, named after Hades, the Greek god of the Underworld, this zone extends from 6,000 to 10,000 m (20,000 to 33,000 ft).

isopod
An order of invertebrates belonging to the crustacean group and including both terrestrial and aquatic species, all having bodies with seven pairs of legs

marine snow
A shower of organic material or biological debris that falls from upper waters to the deep ocean.

mesopelagic
Of or relating to depths between approximately 150 and 1,500 m (500 and 5,000 ft).

mollusc
Any invertebrate with a soft body lacking segments, usually enclosed in a calcareous shell.

mutualism
A mutually beneficial association between two different species.

nekton
Aquatic animals that can actively propel themselves through the water column independent of currents.

nematocysts
Microscopic spheres containing a coiled, barbed filament that is discharged when catching prey or defending against enemies.

nerite
Any member of the family Neritidae, which constitutes small to medium-sized saltwater and freshwater snails.

neuston
Small aquatic organisms found inhabiting the surface layer or attached to the surface film of water.

ooids
Sand-sized ovoidal grains of calcium carbonate formed around a central nucleus found in marine or lacustrine (lake) environments.

pelagic
Of or relating to the open sea or ocean.

photon
A minute energy packet of electromagnetic radiation, and the smallest possible particle of light.

photophore
A light-producing organ present in bioluminescent fish and other bioluminescent organisms.

phylum
A scientific term grouping together related organisms that share fundamental characteristics.

phytoplankton
Often microscopic plants that float in the upper part of the ocean, fuel their metabolism by photosynthesis, and are the base of aquatic food webs. Compare zooplankton.

plankton
Often microscopic plant and animal marine drifters, carried along by tide and currents.

plate tectonics
A geological theory explaining the movement of Earth's lithosphere (crust and upper part of mantle).

polychaete worms
A diverse and abundant group of segmented worms that live in nearly all marine habitats.

polyp
A type of simple animal body form characterized by a more or less fixed base, columnar body and free end with mouth and tentacles, found in various marine organisms, particularly among, for example, corals and jellyfish.

primary production
A biological process to convert light energy, typically from the sun, into chemical energy stored in molecules of glucose.

productive waters
Aquatic environments that have high biological productivity, sufficient to support large amounts of plant and animal life thanks to an abundance of nutrients and favourable conditions.

protozoan
A diverse group of microscopic single-celled organisms.

radiolarian
A group of single-celled marine protozoans that are known for their complex and beautiful silica skeletons.

setae
Hairlike bristles on invertebrates.

symbiont
An organism that lives in a close relationship with another organism.

symbiosis
A close, long-term biological interaction between two or more different biological species, where one or all parties benefit.

Tethys Ocean
A vast and ancient ocean that existed 250 million to around 50 million years ago.

ventral
Belonging to or on or near the lower or abdominal plane of the body.

zooplankton
Very small animals that float in the upper part of the ocean and feed on other organisms. Compare phytoplankton.

zooxanthellae
Tiny plant-like organisms that live in the tissues of many animals such as corals and sponges.

Further Reading

Rewilding the Sea: How to Save our Oceans – *by Charles Clover (London, 2023)*
Clover's globetrotting book is full of hopeful stories of how the ocean can recover from human impacts and be rejuvenated, given a chance.

The Bathysphere Book: Effects of the Luminous Ocean Depths – *by Brad Fox (New York and London, 2023)*
In the 1930s William Beebe was the first scientist to dive into the twilight zone. He went there inside a metal diving device called the Bathysphere. Fox's book is a captivating exploration of Beebe's dives, the world he saw and the people he worked with.

The Draw of the Sea – *by Wyl Menmuir (London and Minneapolis, 2022)*
Menmuir tells the stories of people with connections to the sea, from surfers and sailors to beachcombers and rock poolers.

'Pleuston of the Pacific Ocean', in L. A. Zenkevoch, Biology of the Pacific Ocean – *by A. I. Savilov (US Naval Oceanographic Office, 1969)*
pp. 1–435. Although largely forgotten, this book is a foundational text for the study of neuston.

The Brilliant Abyss – *by Helen Scales (New York, 2021)*
Scales take you to the deepest parts of our ocean, parts that most will never see, and brings it all to life, while showing us how protecting it rather than exploiting it will benefit humankind.

Why Sharks Matter: A Deep Dive with the World's Most Misunderstood Predator – *by David Shiffman (Baltimore, 2022)*
The social media superstar and shark biologist David Shiffman gives an accessible and in-depth guide to the world of sharks, why they matter for the ocean and what can be done to help their endangered populations recover.

The Outlaw Ocean: Journeys Across the Last Untamed Frontier – *by Ian Urbina (New York and London, 2019)*
Urbina expertly takes us on a gripping adventure that brings into view, for the first time, the lawlessness of the vast and wild ocean, a space where criminality and exploitation are rife because it is far beyond our view.

Below the Edge of Darkness: A Memoir – *by Edith Widder (New York, London and Melbourne, 2021)*
Edith Widder's pioneering efforts to take us to the deep ocean and understand the magic of bioluminescence are more than just a scientific story but very much a personal one, brimming with grit and optimism.

Marine Neustonology – *Y. P. Zaitsev (1971)*
Although largely forgotten, this is one of the only academic books on the topic of the ocean's surface ecosystem.

Author Biographies

Asha de Vos

Asha de Vos is an internationally acclaimed Sri Lankan marine biologist, ocean educator, pioneer of long-term blue whale research within the Northern Indian Ocean, Sri Lanka's first deep-sea explorer, and a strong advocate for diversity and equity in marine conservation. She has degrees from the University of St Andrews, the University of Oxford and the University of Western Australia, but escaped academia to establish her own Sri Lankan-grown non-profit association, Oceanswell – Sri Lanka's first marine conservation research and education organization. De Vos and her work have been showcased internationally by a range of media outlets including the BBC, the *New York Times*, TED and *National Geographic*. In 2018 she was included in the BBC's 100 Women list of most inspiring and influential women worldwide, and named *Lanka Monthly Digest*'s Sri Lankan of the Year. In 2020 she was awarded an inaugural Maxwell/Hanrahan Award in Field Biology while also being named *Scuba Diving Magazine*'s Sea Hero of the Year. In 2021 she was awarded a Vanithaabimani Lifetime Achievement Award and the Tällberg-SNF-Eliasson Global Leadership Prize. In 2023 she was named the equity and diversity champion of the British Ecological Society and an Osher Fellow at the California Academy of Sciences. In 2024, de Vos was appointed to the UN Secretary General's seven-member Scientific Advisory Board. She is also the author of *Humpback Whale: A First Field Guide to the Singing Giant of the Ocean*, part of the Young Zoologist series.

Rebecca Helm

Rebecca Helm is a marine biologist studying life in the open ocean and at the ocean's surface. Helm completed her undergraduate degree in Marine Science at Eckerd College, Florida, before conducting research on jellyfish life cycles as a Fullbright Fellow in South Africa. She received a PhD in Ecology and Evolutionary Biology from Brown University. Helm then conducted research on circadian rhythms as a postdoc at Woods Hole Oceanographic Institution, Massachusetts, and later at the Smithsonian National Museum of Natural History as an NSF Postdoctoral Fellow, where she continued her research on coastal and open-ocean jellies. From 2018 to 2022, Helm was an Assistant Professor at the University of North Carolina Asheville, where she led multiple interdisciplinary projects on the biology of life on the high seas. Helm is now an Assistant Professor in the Earth Commons Institute at Georgetown University.

Anthony (Tony) J. Martin

Anthony (Tony) J. Martin is a Professor of Practice at Emory University in Atlanta, Georgia, where he has taught classes in the environmental sciences for more than thirty years. Martin is a geologist and palaeontologist who studies modern and fossil animal traces, and how life has changed environments through time. He is the author of nine books, including two about Georgia coast environments – *Life Traces of the Georgia Coast* and *Tracking the Golden Isles* – as well as popular books about traces and trace fossils, such as *Dinosaurs Without Bones*, *The Evolution Underground* and *Life Sculpted*. Martin also developed and taught an online Coursera class on extinctions (Extinctions: Past, Present, and Future), co-taught a Great Courses class (Major Transitions in Evolution), and is a popular public speaker. He is a Fellow in The Explorers Club and the Geological Society of America.

Helen Scales

Helen Scales is a marine biologist, acclaimed author and broadcaster who explores the wonders and plight of the oceans and the living planet. Her books, including *What the Wild Sea Can Be* and *The Brilliant Abyss*, have been adapted for stage and screen, and translated into fifteen languages. Her titles for young readers include the global bestseller *What a Shell Can Tell* and the *Scientists in the Wild* series. She writes for the *Guardian* and *National Geographic* magazine, teaches at Cambridge University and is a storytelling ambassador for the Save Our Seas Foundation. Scales divides her time between Cambridge, UK, and the wild Atlantic coast of France.

Andrew D. Thaler

Andrew D. Thaler is a deep-sea ecologist, marine technologist and ocean educator. His work focuses on illuminating how humans used technology to explore and exploit the oceans, from coastal estuaries to the deepest trenches. He develops low-cost tools that allow ocean knowledge seekers around the world to study, understand and protect their local waters. He earned a PhD in Marine Science and Conservation from Duke University, North Carolina.

Peter Godfrey-Smith

Peter Godfrey-Smith is Professor of History and Philosophy of Science at the University of Sydney, and works primarily in the philosophy of biology and the philosophy of mind. His books include *Darwinian Populations and Natural Selection* (2009), which won the 2010 Lakatos Award; *Theory and Reality: An Introduction to the Philosophy of Science* (2003, 2021); and *Other Minds: The Octopus and the Evolution of Intelligent Life* (2016), which has been published in over twenty languages. His most recent is *Living on Earth: Forests, Corals, Consciousness and the Making of the World* (2023).

Author Acknowledgments

As you sit comfortably browsing these pages, from front to back, back to front or simply opening the book randomly and having a read, know that it would not have come to life without the support of several people. I especially appreciate Helen Fanthorpe, my senior editor at Thames & Hudson, who was an immense pleasure to work with. Thanks, Helen, for making this adventure so much easier and certainly a whole lot more fun. Thank you also to the rest of my team at Thames & Hudson, Fleur Jones, Lucas Dietrich, Kate Thomas and Matt Watson-Young, and a shout-out to Here Design, for bringing the words on the page to life. High five also to Adrian Sington, my agent from Kruger Cowne, for doing all the stuff I didn't enjoy while I kept working on this project.

A big thank you to my team of co-authors from the shore to the deepest parts of our oceans: Rebecca Helm, Anthony (Tony) Martin, Helen Scales and Andrew Thaler. Thanks for enthusiastically accepting my invitation, for bringing your expertise in science and storytelling to the table, and for your efforts to make more people fall in love with our oceans.

Last but never least, I would like to thank my family. My parents, Ashley and Fatma, said, at a very young age, 'Do what you love, and you will do it well.' These words give me the confidence to swim against the tide and the courage to stretch myself beyond my imaginings. Writing this book is a result of that. Thank you to Charith for silently and patiently supporting me on my dream adventure.

It is with great joy that I dedicate this book to the world's greatest nephew, Niam, for giving me even more reason to work for and look after the largest part of our planet, and to our ocean, without which I would be nothing.

Index

Page numbers in *italics* refer to illustrations

Picture Credits

2 Magnus Lundgren/Wild Wonders of China; 4a ian west/Alamy Stock Photo; 4c IOSPHOTO/Alamy Stock Photo; 4b Martin Strmiska/Alamy Stock Photo; 5a WaterFrame/Alamy Stock Photo; 5c David Shale/Nature Picture Library; 5b Huw Thomas/Alamy Stock Photo; 6a Nature Picture Library/Alamy Stock Photo; 6b Dennis Frates/Alamy Stock Photo; 7a mauritius images GmbH/Alamy Stock Photo; 9cl David Fleetham/Alamy Stock Photo; 9ar Oliver Thompson-Holmes/Alamy Stock Photo; 9cr SeaTops/Alamy Stock Photo; 9br Daniel Lamborn/Shutterstock; 11 Masa Ushioda/Alamy Stock Photo; 12–3 Blue Planet Archive/Alamy Stock Photo; 14–5 Valery Vishnevsky/Alamy Stock Photo; 16 structuresxx/Shutterstock; 18–19 Terry Donnelly/Alamy Stock Photo; 20l Cath Evans/Alamy Stock Photo; 20r Westend61 GmbH/Alamy Stock Photo; 21a sondem/Alamy Stock Photo; 22–3 Hey Emma Kate/Shutterstock; 24a john briscoe/Alamy Stock Photo; 24b Dennis Frates/Alamy Stock Photo; 25a hanohikirf/Alamy Stock Photo; 25b Mikel Bilbao Gorostiaga-Nature & Landscapes/Alamy Stock Photo; 27 Nikita M production/Shutterstock; 28–9 Based on *Tide Chart Infographic* by Kelvin Tow; 30–31a Tamar Dundua/Alamy Stock Photo; 30–31b Tengku Mohd Yusof/Alamy Stock Photo; 32a Prisma by Dukas Presseagentur GmbH/Alamy Stock Photo; 32c Frank Tozier/Alamy Stock Photo; 32b John Crux/Alamy Stock Photo; 33al Dennis Frates/Alamy Stock Photo; 33ar mauritius images GmbH/Alamy Stock Photo; 33cl Stock for you/Shutterstock; 33cr Guna Ludborza/Shutterstock; 33br Gary Chapman/Alamy Stock Photo; 34–5 Stock for you/Shutterstock; 36bl Marion Bull/Alamy Stock Photo; 36br, 37al Siim Sepp/Alamy Stock Photo; 37ar Marek Uliasz/Alamy Stock Photo; 37cl Photo Resource Hawaii/Alamy Stock Photo; 37c Siim Sepp/Alamy Stock Photo; 37cr, 37bl Susan E. Degginger/Alamy Stock Photo; 37br Backyard Productions/Alamy Stock Photo; 38–9 Based on *The Mangrove Ecosystem: Extreme Conditions and Extremely High Biodiversity* by National Geographic Society; 40–1 Mikel Bilbao Gorostiaga-Nature & Landscapes/Alamy Stock Photo; 41al Visions of America, LLC/Alamy Stock Photo; 41r Mikel Bilbao Gorostiaga-Nature & Landscapes/Alamy Stock Photo; 43 Kertu Saarits/Alamy Stock Photo; 44al Hazel McAllister/Alamy Stock Photo; 44bl Alison Wright/Alamy Stock Photo; 45al Michael

Lidski/Alamy Stock Photo; 45ar Duncan McLachlan/Alamy Stock Photo; 45cl Michalakis Ppalis/Shutterstock; 45b antoniolainezph/Shutterstock; 46 © Laurent Ballesta. @laurentballesta, laurentballesta.com; 48 FLPA/Alamy Stock Photo; 49al mark higgins/Shutterstock; 49ar Steve Taylor ARPS/Alamy Stock Photo; 49bl Lee Rentz/Alamy Stock Photo; 49br Photimageon/Alamy Stock Photo; 50 row 1 from above, left to right: nattawut lakjit/Alamy Stock Photo, PjrShells/Alamy Stock Photo, 50 row 2 from above, left to right: PjrShells/Alamy Stock Photo, Constantinos Zorbas/Alamy Stock Photo, Anastasiia Skorobogatova/Alamy Stock Photo; 50 row 3 from above, left to right: Vitaliy Pakhnyushchyy/Alamy Stock Photo, PjrShells/Alamy Stock Photo, Yongkiet Jitwattanatam/Alamy Stock Photo; 50 row 4 from above, left to right: EpicStockMedia/Alamy Stock Photo, Michael Svetbird/Alamy Stock Photo, Zee/Alamy Stock Photo; 51a Hihitetlin/Alamy Stock Photo; 51al WhiteJack/Alamy Stock Photo; 51ar Vladislav Gajic/Alamy Stock Photo; 51c YAY Media AS/Alamy Stock Photo; 51bl Andrey Elkin/Alamy Stock Photo; 51br Lefteris Papaulakis/Alamy Stock Photo; 52l Maximilian Weinzierl/Alamy Stock Photo; 52r Arterra Picture Library/Alamy Stock Photo; 53 All Canada Photos/Alamy Stock Photo; 56–7 IrinaK/Shutterstock; 58 Andriy Nekrasov/Shutterstock; 60–1 Stephen Frink Collection/Alamy Stock Photo; 62 WaterFrame/Alamy Stock Photo; 63a Alex Mustard/Nature Picture Library; 63b Salty filters/Shutterstock; 64–5 Nature Picture Library/Alamy Stock Photo; 66–7b Nature Picture Library/Alamy Stock Photo; 67a Derek D. Galon/Shutterstock; 67r Tamara Kulikova/Shutterstock; 68 Nature Picture Library/Alamy Stock Photo; 69 Based on *Morphology and development of the Portuguese man of war, Physalia physalis* by Catriona Munro, Zer Vue, Richard R. Behringer and Casey W. Dunn; 70 Jeff Milisen/Alamy Stock Photo; 71a Tony Wu/Nature Picture Library; 71b Hiroya Minakuchi/Nature Picture Library; 72a Chris Johnson/Alamy Stock Photo; 72bl imageBROKER.com GmbH & Co. KG/Alamy Stock Photo; 72br Nature Picture Library/Alamy Stock Photo; 73a BIOSPHOTO/Alamy Stock Photo; 73bl WaterFrame/Alamy Stock Photo; 73cr Stocktrek Images, Inc./Alamy Stock Photo; 73br Nature Picture Library/Alamy Stock Photo; 74–5 Jeff Milisen/Alamy Stock Photo; 76a Solvin Zankl/Nature Picture Library; 76b IrinaK/Shutterstock; 77a © Denis Riek; 77c Andrea Izzotti/Shutterstock; 77b Luis Quinta/Nature Picture Library; 78l ggw/Shutterstock; 78r Ekky Ilham/Shutterstock; 79 Natural Visions/Alamy Stock Photo; 80–1 Based on *Summary of key adaptations of Halobates for oceanic life* by Xavier Pita, Senior Scientific Illustrator, KAUST; 82 Photo Dr. Leonid Svetlichny; 83a B.A.E. Inc./Alamy Stock Photo; 83bl Bazzano Photography/Alamy Stock Photo; 83br Anthony Pierce/Alamy Stock Photo; 84a

blickwinkel/Alamy Stock Photo; 84b Erin Donalson/Alamy Stock Photo; 85a Robert Wallwork/Alamy Stock Photo; 85c ANESTIS REKKAS/Alamy Stock Photo; 85b Blue Planet Archive/Alamy Stock Photo; 86 Sabena Jane Blackbird/Alamy Stock Photo; 87al Martin Shields/Alamy Stock Photo; 87ar Hani Amir/Shutterstock; 87bl Science Photo Library/Alamy Stock Photo; 87br Ivan Vdovin/Alamy Stock Photo; 89 Based on a figure from *Why did only one genus of insects, Halobates, take to the high seas?* by Lanna Cheng and Himanshu Mishra; 90a Solvin Zankl/Nature Picture Library; 90c David Chapman/Alamy Stock Photo; 90b All Canada Photos/Alamy Stock Photo; 91 Nature Photographers Ltd/Alamy Stock Photo; 92–3 Pascal Kobeh/Nature Picture Library; 94–5 Gary Bell/Oceanwide/Nature Picture Library; 96 WaterFrame/Alamy Stock Photo; 98–9 The Metropolitan Museum of Art, New York. Gift of Mr. and Mrs. Erving Wolf, in memory of Diane R. Wolf, 1977; 100al RLS PHOTO/Alamy Stock Photo; 100–1 WaterFrame/Alamy Stock Photo; 100bl Victor Savushkin/Alamy Stock Photo; 101a Connect Images/Alamy Stock Photo; 102–3 Martin Strmiska/Alamy Stock Photo; 104al Nature Picture Library/Alamy Stock Photo; 104ac, 104ar, 104cl Ruben Duro/Science Photo Library; 104cra Helmut Corneli/Alamy Stock Photo; 104crb Photo Solvin Zankl/Nature Picture Library; 104bl blickwinkel/Alamy Stock Photo; 104bc Blue Planet Archive/Alamy Stock Photo; 104br Solvin Zankl/Nature Picture Library; 105 from above to below, left to right: mauritius images GmbH/Alamy Stock Photo, Solvin Zankl/Nature Picture Library, Nature Picture Library/Alamy Stock Photo, FLPA/Alamy Stock Photo, Ingo Arndt/Nature Picture Library, Blue Planet Archive/Alamy Stock Photo, Wim Van Egmond/Science Photo Library, WaterFrame/Alamy Stock Photo, Ingo Arndt/Nature Picture Library, Sinclair Stammers/Nature Picture Library; 106–7 Based on *The Importance of Seagrass* by Wildlife Conservation Society, New York; 108a SeaTops/Alamy Stock Photo; 108bl Jason Doucette/Alamy Stock Photo; 108br, 109a Marli Wakeling/Alamy Stock Photo; 109c imageBROKER.com GmbH & Co. KG/Alamy Stock Photo; 109bl SeaTops/Alamy Stock Photo; 109br imageBROKER.com GmbH & Co. KG/Alamy Stock Photo; 110–1 Daniel Lamborn/Alamy Stock Photo; 112al Christopher D. Morgan/Shutterstock; 112ar cbimages/Alamy Stock Photo; 112cl blickwinkel/Alamy Stock Photo; 112cr Henner Damke/Alamy Stock Photo; 112bl WaterFrame/Alamy Stock Photo; 112br imageBROKER.com GmbH & Co. KG/Alamy Stock Photo; 113al Hubert Yann/Alamy Stock Photo; 113ar imageBROKER.com GmbH & Co. KG/Alamy Stock Photo; 113cl 22August/Shutterstock; 113cr Reinhard Dirscherl/Alamy Stock Photo; 113bl MadeleinWolf/Alamy Stock Photo; 113br imageBROKER.com GmbH & Co. KG/Alamy Stock Photo; 114–5a Rod Haestier/Alamy

Stock Photo; 114–5b Jerome Murray – CC/Alamy Stock Photo; 118–9 Matthew Oldfield Underwater Photography/Alamy Stock Photo; 120al Oliver Thompson-Holmes/Alamy Stock Photo; 120ar Andrey Nekrasov/Alamy Stock Photo; 120bl, 120br Oliver Thompson-Holmes/Alamy Stock Photo; 121al Joe Belanger/Alamy Stock Photo; 121ar Jeff Rotman/Alamy Stock Photo; 121bl Helmut Göthel Symbiosis/Alamy Stock Photo; 121bc, 121br Oliver Thompson-Holmes/Alamy Stock Photo; 122–3 WaterFrame/Alamy Stock Photo; 124 © Michael Deal; 125a, 125bl © 1971, The American Association for the Advancement of Science; 125bc Courtesy Ocean Alliance; 125br ruelleruelle/Alamy Stock Photo; 128al WaterFrame/Alamy Stock Photo; 128ar imageBROKER.com GmbH & Co. KG/Alamy Stock Photo; 128b Helmut Corneli/Alamy Stock Photo; 128–9 James/Alamy Stock Photo; 129a Georgette Apol/Alamy Stock Photo; 130, 131 Ernst Heinrich Philipp Haeckel, *Art Forms of Nature (Kunstformen der Natur),* 1904. Leipzig und Wien: Verlag des Bibliographischen Instituts; 132–3 Based on *Whale phylogeny from The Tangled Bank* by Carl Zimmer; 134–5 Based on *Evolution of fishes from the Ordovician period to the present day* by Gwen Shockey/Science Source; 136–7 Panther Media GmbH/Alamy Stock Photo; 138al Stocktrek Images, Inc./Alamy Stock Photo; 138ar ullstein bild/Getty Images; 138b Aflo/Nature Picture Library; 140–1 Carl Chun, *Scientific results of the German deep sea expedition on the steamer "Valdivia",* 1898–1899; 142 © Dr. J.Mallefet – FNRS; 143l WaterFrame/Alamy Stock Photo; 143r Steve Trewhella/Alamy Stock Photo; 144–5 Nature Picture Library/Alamy Stock Photo; 146, 147 Frank Thomas Bullen, *Creatures of the Sea: Being the Life Stories of Some Sea Birds, Beasts, and Fishes,* 1857–1915; 148al Photo Larry Madin. Woods Hole Oceanographic Institution; 148ar Monterey Bay Aquarium Research Institute (MBARI); 148bl David Fleetham/Alamy Stock Photo; 148br Jeff Rotman/Alamy Stock Photo; 149a Monterey Bay Aquarium Research Institute (MBARI); 149bl Yiming Chen/Getty Images;149br © MBARI/NOAA; 150, 151 William Beebe and Ruth Rose, *The Arcturus adventure: an account of the New York Zoological Society's first oceanographic expedition,* 1926; 152al Photo Sönke Johnsen; 152ar Nature Picture Library/Alamy Stock Photo; 152bl © Woods Hole Oceanographic Institution; 152br Monterey Bay Aquarium Research Institute (MBARI); 153a David Shale/Nature Picture Library; 153c © Woods Hole Oceanographic Institution; 153b Solvin Zankl/Nature Picture Library; 154–5 Based on *Deep pelagic food web structure as revealed by in situ feeding observations,* by C. Anela Choy, Steven H. D. Haddock and Bruce H. Robison; 154–5 from above to below, left to right: mauritius images GmbH/Alamy Stock Photo, Nature Photographers Ltd/Alamy Stock

Photo, SeaTops/Alamy Stock Photo, Blue Planet Archive/Alamy Stock Photo, WaterFrame/Alamy Stock Photo, Stocktrek Images, Inc./Alamy Stock Photo, Nature Picture Library/Alamy Stock Photo, Monterey Bay Aquarium Research Institute (MBARI), Solvin Zankl/Nature Picture Library, David Shale/Nature Picture Library, Colin Marshall/Alamy Stock Photo, David Shale/Nature Picture Library, David Shale/Nature Picture Library, Nature Picture Library/Alamy Stock Photo, Andrey Nekrasov/Alamy Stock Photo, Monterey Bay Aquarium Research Institute (MBARI), Shutterstock, Photo Fabio Vitale. Courtesy Salento Sommerso, Nature Picture Library/Alamy Stock Photo, Solvin Zankl/Nature Picture Library; 156 Aleksandar Milutinovic/Alamy Stock Photo; 157 Henk-Jan Hoving, GEOMAR; 160–1 Adisha Pramod/Alamy Stock Photo; 162, 163 A. Giltsch, E. Haeckel, *Hexacoralla - Sechsstrahlige Sternkorallen,* 1904. Library of Congress. Prints and Photographs Division, Washington, D.C.; 164a Georgette Apol/Alamy Stock Photo; 164–5 Solvin Zankl/Nature Picture Library; 165al E.R. Degginger/Alamy Stock Photo; 165ar blickwinkel/Alamy Stock Photo; 165br Fearnstock/Alamy Stock Photo; 166–7 Fundação Rebikoff-Niggeler; 168al, 168ar David Shale/Nature Picture Library; 168–9 Kate Thomas. Duke University; 168bl Nature Picture Library/Alamy Stock Photo 168br NOAA/Alamy Stock Photo; 169c World History Archive/Alamy Stock Photo; 169bl, 169br David Shale/Nature Picture Library; 170 The Natural History Museum/Alamy Stock Photo; 171a © Laurent Ballesta. @laurentballesta, laurentballesta.com; 171b John Cancalosi/Nature Picture Library; 172–3 from above, left to right: David Shale/Nature Picture Library, Nature Photographers Ltd/Alamy Stock Photo, Monterey Bay Aquarium Research Institute (MBARI), Yiming Chen/Getty Images, David Shale/Nature Picture Library, Peter Bucktrout/Bas/Shutterstock; 174 Sunny Celeste/imageBROKER/Shutterstock; 175 Brandon Cole/Nature Picture Library; 176a WaterFrame/Alamy Stock Photo; 176b SeaTops/Alamy Stock Photo; 177al David Shale/Nature Picture Library; 177ar WaterFrame/Alamy Stock Photo; 177c Juergen Freund/Nature Picture Library; 177bl Monterey Bay Aquarium Research Institute (MBARI); 177br Stocktrek Images, Inc./Alamy Stock Photo; 178 © Purix Verlag Volker Christen/Bridgeman Images; 179 F. McCoy, *Echinorhinus brucus, Natural history of Victoria. Prodromus of the zoology of Victoria,* 1885; 182–3 Solvin Zankl/Nature Picture Library; 184–5 Sharon Eisenzopf/Shutterstock; 186 The Natural History Museum/Alamy Stock Photo; 188–9 Ocean Exploration Trust/NOAA; 190a NOAA Okeanos Explorer Program, INDEX-SATAL 2010, NOAA/OER; 190bl WILDLIFE GmbH/Alamy Stock Photo; 190br David Hall/Nature Picture Library; 191a Ocean Exploration Trust/NOAA; 192–3 BIOSPHOTO/

Alamy Stock Photo; 194, 195al NOAA/Monterey Bay Aquarium Research Institute; 195ar Deepwater Canyons 2013-Pathways to the Abyss, NOAA-OER/BOEM/USGS; 195cl NOAA NMFS SWFSC Antarctic Marine Living Resources (AMLR) Program; 195cr Schmidt Ocean Institute; 195bl Courtesy Expedition to the Deep Slope 2007, NOAA-OE; 195br NOAA/Monterey Bay Aquarium Research Institute; 196–7 David Shale/Nature Picture Library; 198–9 Brandon Cole/Nature Picture Library; 199a 19th era 2/Alamy Stock Photo; 199b Joel Sartore/Photo Ark/Nature Picture Library; 200 David Shale/Nature Picture Library; 201 Fifis Alexis, Ifremer (2005); 202–3 © Armando Veve; 204–5 WaterFrame/Alamy Stock Photo; 208–9 Nature Picture Library/Alamy Stock Photo; 210–1 Lamont-Doherty Earth Observatory and the estate of Marie Tharp; 211a AIP Emilio Segrè Visual Archives. Gift of Bill Woodward. USNS Kane Collection; 211r Lamont-Doherty Earth Observatory and the estate of Marie Tharp; 212–3 NOAA/Monterey Bay Aquarium Research Institute; 214al David Shale/Nature Picture Library; 214–5 RLS PHOTO/Alamy Stock Photo; 215a Tony Wu/Nature Picture Library; 216–7 SeaTops/Alamy Stock Photo; 218–9 Based on *Lakes and Oceans* by Randall Munroe; 220, 221 Ocean Exploration Trust/NOAA; 222a Courtesy Submarine Ring of Fire 2006 Exploration, NOAA Vents Program; 222b Ocean Exploration Trust/NOAA; 224–5 Science History Images/Alamy Stock Photo; 225a Keystone Press/Alamy Stock Photo; 225r Library of Congress. Rare Book and Special Collections Division Washington, D.C.; 226–7 Based on *The processes involved in deep-sea mining for the three main types of mineral deposits* by Miller et al. 2018; 228–9 Adisha Pramod/Alamy Stock Photo; 230 imageBROKER.com GmbH & Co. KG/Alamy Stock Photo; 231 Contraband Collection/Alamy Stock Photo; 232–3 Google Maps/Google Earth; 234–5 WaterFrame/Alamy Stock Photo; 236–7 Imaginechina Limited/Alamy Stock Photo; 238 Imaginechina Limited/Alamy Stock Photo; 240–1 Michael Fritzen/Alamy Stock Photo; 242–3 domonabikeSingapore/Alamy Stock Photo; 243 © National Maritime Museum, Greenwich, London; 244 Bettmann/Getty Images; 245al Florilegius/Alamy Stock Photo; 245ar ullstein bild/Getty Images; 245c Stock Italia/Alamy Stock Photo; 245bl ClassicStock/Alamy Stock Photo; 245br API/Gamma-Rapho/Getty Images; 246–7 Anwaruddin Tajudin/Alamy Stock Photo; 248–9a Timothy Allen/Getty Images; 248–9b Imaginechina Limited/Alamy Stock Photo; 250–1 Matjaz Corel/Alamy Stock Photo; 251a Kim Wonkook/Shutterstock; 251b Various images/Shutterstock; 252a AF Fotografie/Alamy Stock Photo; 252bl Library of Congress. Prints and Photographs Division, Washington, D.C.; 252br Glasshouse Images/Alamy Stock Photo; 253a The Metropolitan Museum of Art, New York. Harris Brisbane Dick

First published in the United Kingdom in 2025 by
Thames & Hudson Ltd, 6–24 Britannia Street, London WC1X 9JD

First published in the United States of America in 2025 by
Thames & Hudson Inc., 500 Fifth Avenue, New York,
New York 10110

EU Authorized Representative: Interart S.A.R.L.
19 rue Charles Auray, 93500 Pantin, Paris, France
productsafety@thameshudson.co.uk
interart.fr

A CIP catalogue record for this book is available from the
British Library

Library of Congress Control Number 2025930082

ISBN 978-0-500-02755-4
01

Printed and bound in China by Toppan Leefung Printing Limited.

Be the first to know about our new releases,
exclusive content and author events by visiting
thamesandhudson.com
thamesandhudsonusa.com
thamesandhudson.com.au

The infographics in this book are not to scale.

Please see below for captions to images on the following pages:

p. 2: Oval squid (*Sepioteuthis lessoniana*)

pp. 4–5 (clockwise from top left): Ammonite fossil, Lyme Regis, Dorset, UK; Coloured squat lobster (*Trapezionida olivarae*), Kai Islands, Moluccas, Indonesia; Snailfish (Liparidae), Mid-Atlantic Ridge, North Atlantic Ocean; Scuba divers exploring a tropical reef; Coral reef, Komodo National Park, Indonesia; Portuguese man o' war (*Physalia physalis*), Tenerife, Canary Islands, Spain

pp. 12–13: Sharpear enope squid (*Ancistrocheirus lesueurii*), Kona Coast, Big Island, Hawaii, USA

p. 16: Aerial view of coast's natural shape and texture at low tide in Thailand

pp. 22–23: Aerial view of coast with camels and four-wheel-drive vehicles, Cable Beach, Broome, Australia

pp. 34–35: Tidal flats at low tide at Donggeomdo Island near Ganghwa-gun, South Korea

p. 58: *Sargassum* seaweed floats at the air–sea interface and can form large rafts of algae that serve as habitat for diverse species

pp. 64–65: Sargasso weed (*Sargassum fluitans*) with pugnose pipefish (*Syngnathus pelagicus*), Sargasso Sea, Bermuda

pp. 74–75: Blue sea dragons (*Glaucus*) eat a variety of surface species, including Portuguese man o' war (*Physalia physalis*) (seen here)

pp. 92–93: Giant barnacles (*Balanus nubilus*), Gulf of Alaska, USA

p. 96: Colourful reef of fan corals (*Melithaea*)

pp. 102–103: Colourful coral reef scene with purple soft corals and tropical fish, Safaga, Red Sea, Egypt

pp. 110–111: Sardine run

pp. 118–119: Nudibranch, Lembeh Strait, Sulawesi, Indonesia

pp. 122–123: Teeth of bridled parrotfish (*Scarus frenatus*)

p. 138: Clockwise from top left: Bigfin reef squid; squid; fish larva

pp. 144–145: Viperfish (*Chauliodus sloani*)

pp. 166–167: A female fanfin angler (*Caulophryne jordani*) with a tiny male attached underneath

pp. 182–183: Glass octopus (*Vitreledonella richardi*)

p. 186: The bone-eating snot-flower worm (*Osedax mucofloris*)

pp. 192–193: Blotched snailfish (*Crystallichthys cyclospilus*) on reef, Gambier Bay, Alaska, USA

pp. 196–197: Scaly-foot gastropod (*Crysomallon squamiferum*)

pp. 200–201: Different views of yeti crabs (*Kiwa hirsuta*)

pp. 208–209: Giant deepsea isopod (*Bathynomus giganteus*)

pp. 212–213: A Venus flytrap anemone (*Actinoscyphia*) attached to the deep-sea floor

pp. 228–229: The bone-eating snot-flower worm (*Osedax mucofloris*)

pp. 234–235: Coldwater soft coral (*Gersemia fruticosa*), White Sea, Karelia, Russia

p. 238: Bajau children diving in Semporna, Malaysia

pp. 246–247: Aerial view of Bajau homes at Tatagan Island, Semporna, Malaysia

pp. 254–255: Peruvian pelican (*Pelecanus thagus*) feeding frenzy, Pucusana, Peru (Humboldt Current)

pp. 272–273: Scuba diver exploring a coral reef in the Red Sea, Egypt